T0384075

BALANCING A SAUROPOD

BALANCING A SAUROPOD

The Physiology of a Dinosaur

BRANT E. ISAKSON

Professor of Molecular Physiology and Biological Physics, Robert M. Berne Cardiovascular Research Center, University of Virginia School of Medicine, Charlottesville, VA, United States

Academic Press is an imprint of Elsevier
125 London Wall, London EC2Y 5AS, United Kingdom
525 B Street, Suite 1650, San Diego, CA 92101, United States
50 Hampshire Street, 5th Floor, Cambridge, MA 02139, United States

ISBN 978-0-12-823303-0

For information on all Academic Press publications
visit our website at https://www.elsevier.com/books-and-journals

Publisher: Mara Conner
Acquisitions Editor: Elizabeth Brown
Editorial Project Manager: Susan Ikeda
Production Project Manager: Gomathi Sugumar
Cover Designer: Vicky Pearson Esser

Typeset by STRAIVE, India

Dedication

To my wife Amy, who is all things.
To my dad, and the memory of my mom.
And to my boys, Nick, Joe, and Tommy; they think I'm a dinosaur.

Contents

Epigraph

There are known knowns. These are things we know that we know. There are known unknowns. That is to say, there are things that we know we don't know. But there are also unknown unknowns. There are things we don't know we don't know.

Donald Rumsfeld

Introduction

Balancing a sauropod might sound like an impossible task—not literally by having them sit atop a fulcrum and hold perfectly still, but instead by understanding how a sauropod's physiological processes achieved balance between all of their cells, tissues, and organs, harmoniously interacting second-to-second to produce a developing, breathing, eating, and reproducing dinosaur with a 50-year life expectancy and, at a macro-level, with the sauropod species finding a way to evolve and thrive as the world around them changes dramatically, with variations in temperature and weather patterns, land mass and oceanic formation, earthquakes, and rapid vegetation evolution.

After years of study, we understand how many of today's species take in oxygen and how that oxygen travels throughout the body, producing the power for necessary anatomical processes and life-sustaining daily activities. But how did the gigantic sauropod take in enough oxygen in an oxygen-deprived environment where they lived, let alone have enough energy to move that enormous tail of theirs (e.g., in response to a scratchy fern or pesky *Tyrannosaurus rex*)?

There are some dinosaur species with a body habitus and size that we can wrap our head around—think of triceratops or even a velociraptor, popularized in movies. But what about something like a titanosaurus? A species of dinosaur belonging to the sauropods family, characterized by their absolutely enormous size, including the top of their head to the tip of their tail, measuring greater than 30 m. That is roughly the length of three buses back-to-back, two volleyball courts, or a 10-story building laying on its size! Now, how does an animal that large in shear size (this aside from the extremely long neck which presents additional challenges) sustain itself to old age?

And not only is the size of the sauropod hard to wrap your mind around, but the questions of how they actually survived and thrived are mind-boggling. The sauropod family of species lived all across the world, in each corner of the globe, for over 150 million years. How could an animal of this size breathe in enough oxygen to live, have enough energy to sustain itself for 50 years, and propagate their species across the globe for that long? These are the questions this book tries to *begin* to answer.

To go about answering these questions, I have attempted to write this book in a way that merges technical components of physiology—what I consider important fundamental knowledge about how certain organ systems work—with some extrapolated ideas about sauropods. I have focused heavily on respiration, vasculature, and the heart—essentially the cardiovascular system. The reason for this is because I believe these three systems are lynchpins for understanding sauropod physiology. Due to their gigantic size, their long neck, and the environment where these sauropods lived, how they were able to extract oxygen from the environment and transport it throughout their body would seem one of the most critical components of their physiological engine and the basis of their existence. I have also included a chapter with kidney function and metabolism, as these physiologies go together with the cardiovascular system. I have not purposefully excluded bone structure, muscles, reproduction, or even possible brains, except that the fossil record is more informative in these instances than with the cardiovascular system, which exists only as hypotheticals.

I have borrowed heavily from comparative animal physiology, especially the giraffe. The reason for this heavy focus was of course the comparable neck lengths, with, in some sauropod species, a comparable distance from the heart to the brain. I would point the reader to Dr. Christian Akjær for an amazingly detailed and extensive knowledge of the physiology of these wonderful animals. For size, I used comparisons to modern-day blue whales, but this comparison is less sparse due to the whale's exclusive water environment and the lack of any meaningful distance between the heart and brain. Although both of these species are mammals, and our friend the sauropod wasn't, there are many aspects of physiology that are highly conserved (e.g., the four-chamber heart), and I have tried to provide justification for why the sauropod may have had similar or different features compared with these mammals.

The last chapter tries to pull together as much as possible from these technical chapters, and present an early, conceptual idea of sauropod physiology. Ultimately, this book will hopefully provide insight into the extremes of physiology and how adaptions can be made to tweak organ systems to keep things humming along in homeostasis, an idea we will continuously return to.

I have made broad strokes in areas of physiology and paleontology (especially here). This is not to dismiss anyone or anything; I am simply trying to get the big picture explained, so that we can use this to try to start understanding sauropod's physiology. Biology is constantly evolving with new discoveries, leading to new interpretations. I'm sure the same will happen here. Maybe there will be soft tissue remains (however doubtful) showing a sauropod's heart that would completely change our understanding of how they worked? Or maybe more likely an outline of air sacs against their rib cage? Whatever the case, this is where we are currently at in terms of the fossil record and how physiology may intersect.

Having said earlier, an important admission for this book is that I am not a paleontologist. So with that acknowledgment, I ask forgiveness from my paleontologist colleagues if I miss a sauropod's name or skip over a seminal paper in the field. I sincerely hope that is not the case, but it is not my training and I have done my best. One important note in that regard is I have essentially lumped all sauropods together into one broad species; I am fully aware there are many completely different sauropod species, each with their own unique size, neck length, and likely ecological niches. I have done this to simplify our discussion. Hopefully, this book, by a card-carrying physiologist, adds a layer to the incredibly hard work done by paleontologists and, at the very least, provides some new food for thought.

For helping push me through, I thank my dear wife, Amy Isakson, for taking the time to read through this entire book, word for word. I want to say thanks to my many colleagues who listened to my wanderings on sauropods, especially my good friend Miriam Cortese-Krott. In addition, I thank the many members of my lab who indulged my interest in sauropods over the years, either in lunchtime conversations or taking up journal club time; these include Dr. Katie Heberlein, Dr. Alex Lohman, Dr. Lauren Biwer, Dr. Leon DeLalio, Dr. T.C. Steven Keller IV, Dr. Nenja Krüger, Dr. Alex S. Keller, Dr. Yang Yang, Dr. Claire A. Ruddiman, Dr. Abby Wolpe, Dr. Melissa A. Luse, Wyatt Schug, Skylar Loeb, Zuzanna Juskiewicz, Edgar Macal, and Shruthi Nyshadham, as well as Dr. Luke Dunaway, Dr. Miranda Good, Dr. Isola Brown, Dr. Scott Johnstone, Dr. Josh Butcher, Dr. Adam Straub, Dr. Marie Billaud, Dr. Henry Askew-Page, Dr. Daniella Begandt, and Dr. Xiahong Shu. Much gratitude to the many, many international students and undergraduates that I have been fortunate to work with. And of course, thanks to my long-time lab manager, Angela K. Best.

Finally, I thank Anita Impagliazzo for the illustrations she drew. All of the illustrations in this book are her original artwork that I feel are an amazing and essential component to the discussion on sauropod's physiology. Her drawings bring to life a lot of the crazy thoughts I had on how a sauropod may have worked—thank you Anita!

CHAPTER ONE

What do we know about the sauropods?

Dinosaurs have long filled stories and imaginations. They have been made into toys, been the inspiration for blockbuster movies, and entertained children through cartoon renderings. This fame is without their species ever intermingling with our own. Their existence is often discussed and postulated by scientists and nonscientists and adults and children alike, with an allure that is as ageless as time itself. And of the many species of dinosaurs, the large group of species broadly defined as sauropods, with their enormous size, long-neck, and pot-belly, is the most well-known and also some of the first dinosaurs discovered.

As initially described by Benoit and Helm in 2023, phalanx of *Massospondylus dinosaurs*, an early sauropod species, was found deliberately placed in an ancient shelter in Lesotho and South Africa, dating back possibly as far as AD 1100. Finding these bones was likely possible because the surrounding geology was full of sediment from nearly 200 million years ago, and rains commonly exposed the massive fossils. These ancient people realized the phalanx was from a huge animal and called them "Kholumolumo" or "Amagongqongpo," imagining them as dragon-like creatures that were big enough to eat houses. They weren't too far off when describing the size of the *Massospondylus*, and serendipitously, it was one of the few carnivorous sauropods!

However, the modern glimmer of discovery that was to be become recognized as sauropods happened in an environment that many sauropod species may have felt at home in—lush and overgrown terrain covered with green fern-like plants. It was 1699 in Wales, United Kingdom, and Edward Lhuhyd was tinkering with some geologically unique rocks along Swansea Bay when he stumbled upon odd-looking teeth that didn't appear to be like anything he could recognize. Little did he know how important those teeth would become in understanding some of the basic biology of the creature he stumbled upon. This was because, at that time, there was no comprehension of dinosaurs, but it was clear the teeth belonged to an animal that was not derived from Wales (but maybe a whale?).

Balancing a Sauropod
https://doi.org/10.1016/B978-0-12-823303-0.00008-2

It took another 200 years before Dr. Othniel Charles Marsh at Yale University in the United States would fully flush out sauropods as a species in the late 1800s, and we would come to recognize that the teeth Lhuhyd discovered were from the sauropod species *Rutellum impicatum*, a gigantic creature about the size of a house. There's no doubt that Lhuhyd would have been astonished if he knew the teeth he discovered belonged to a sauropod dinosaur, one of the most successful and conversely one of the strangest groups of animals to ever live on earth.

Sauropods were worldwide, and although we know them as herbivores, early sauropods were also carnivorous and had a huge array of mechanisms to eat and move, an important component in adapting to the multiple geological and geographical locations across prehistoric earth. The sauropods burst on the evolutionary scene with a swagger, literally and metaphorically, and were altogether different than the dinosaurs that had come before them. They dominated the earth for 160,000,000 years. For comparison, *Homo sapiens* (humans) have been around ~300,000 years (not even 0.2% of the time!).

Indeed, the sauropod species of dinosaurs were still going strong by the time the meteor hit the Yucatan Peninsula of the current-day Mexico and caused the Chicxulub impact 65 million years ago. Thus, the sauropod family of dinosaurs could generally be considered quite a "successful" animal in terms of surviving on earth for an extended period of time. In fact, there are not a lot of known animals, dinosaurs or otherwise, that have this type of success, living in a recognizable form for such a long period of time. It certainly wasn't the same type of sauropod that entire 160 million year span, but there are overarching themes to this family of dinosaurs, their gigantism and long neck being the stereotypical feature. There are of course some interesting exceptions to this, including the *Nigersaurus taqueti*, which was about the size of a large cow, and *Brachytrachelopan* with its very short neck. However, they would all be easily recognized among their sauropod cousins based on their neck position and head placement, large abdomen, and elongated tail.

Standing on the shoulders of giants

In terms of diverse localization and species, dinosaurs have been one of the most successful animals to have colonized the earth, second only to perhaps insects. Dinosaurs thrived for hundreds of million years in multiple sizes, shapes, colors, and locations on earth. However, of all the different arrays of dinosaurs, the line of dinosaurs broadly termed sauropodomorphs stands out the most. It's safe to say that almost anyone who has witnessed a mounted sauropod in a museum has been astonished at its sheer size.

For example, in Berlin's Museum für Naturkunde, you practically walk into a 13 m-tall *Giraffatitan brancai* in the central atrium. It's the centerpiece of the museum and provides a memorable welcome to the facility. In New York City's American Museum of Natural History, the *Patagotitan mayorum* titanosaur can't even fit into the exhibit hall! Even in smaller museums like the University of Wyoming's Geological Museum in Laramie Wyoming, the original *Apatosaurus* sauropod fills the entire room with a toothy skull greeting you as you walk in the door. Although we only have sauropod fossils, paleontologists can estimate their weight with many different sauropod species weighing up to nine elephants. Regardless of this massive size, according to inferences from the fossil record, sauropods were actually quite mobile, rapid, and agile creatures (more on that later). We also can infer several other components to sauropods such as how they walked, how they held their massive body together, and even possibly what they ate.

Perhaps the least well known is how these animals worked as an organism—by that, I mean, for example, how their hearts beat and how they lived in such hypoxic environments. We simply have nothing similar today.

For the closest comparison purposes, we can try to use giraffes and blue whales. Giraffes are only one-fifth the size of a stereotypical sauropod, but it's their elongated necks that make such an important comparison. The distance from the heart to the brain via the neck will be discussed many times, and giraffes are the only living animal that come close to this comparison in sauropods. Other perhaps less-distant species likes the avian emus and ostriches also have long necks, but the distance from the heart to the brain is considerably smaller than that in giraffes, and the weight of the neck and head compared with their body is more in alignment with other species (see Fig. 1.1).

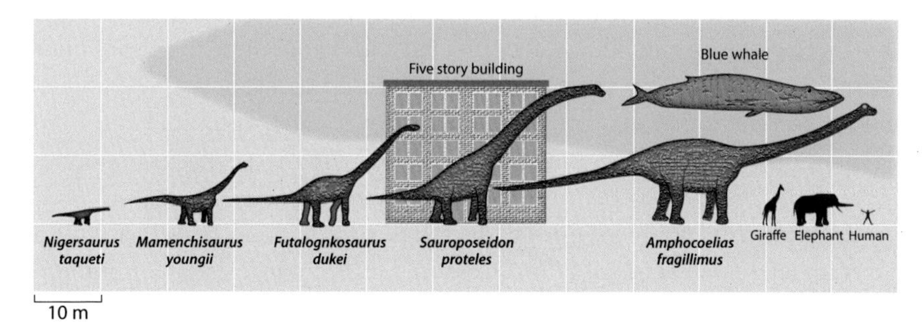

Fig. 1.1 Size comparison of the sauropod.

The closest living species in terms of overall weight and size are blue whales. We can use blue whales to examine the rare feature of gigantism in animals and how this may dictate things, such as endocrine controls or nervous system control over long distances. However, this is also a difficult comparison, as whales live in the ocean, and so the effects of gravity are significantly less to deal with from the cardiovascular system to the weight-bearing bones. Regardless of what we use to compare the sauropods, too, we still lack direct information of how the sauropods functioned. To understand these important details, we must turn to the study of physiology, combining it with what we know from sauropod's paleontology. However, this comparison must be done cautiously, because intact organs governing much of the sauropod physiology (e.g., hearts, stomachs, and brains) is soft tissue and does not survive as dinosaur fossils, or very rarely.

To begin to put together how a sauropod may have functioned, and why it was an excellent example of what we call extreme physiology, we first must break down some of the key parts of what we know about sauropods, starting with understanding what exactly a fossil is.

No bones about it—What is a fossil?

Let's remember for a moment what fossils are: either body fossils or trace fossils—body fossils are from previous animals' organic matter (e.g., bones), and trace fossils are preserved from animal activities and not real body parts (e.g., footprints). Body fossils are actually rare and require a fair amount of luck to form.

For a body fossil to form, the animal or organic organism must be rapidly covered in sediment of some kind (e.g., sand or ash) that is compacted repeatedly. The organic material under the sediment decomposes, and what's left is essentially the mold of where the animal used to be. Water seeps into the sediment mold and deposits minerals that outline where the creature was. So, body fossils aren't actual bones, but a mold of where the bones once were. The reason why the bones, or what used to be bones, are almost always present is that the soft tissue decomposes much faster and is feasted on by the tiny creatures in the soil. Thus, the exquisite details in the bones are one of the best ways to understand sauropods.

As a further side note, under perfect preservation conditions that would be completely independent of fossilization, DNA could last about 6 million years but would not likely be readable by the current technology after about 1.5 million years ago. Thus, the concept of a Jurassic Park, no matter how amazing it could be, is likely never possible (unfortunately!). We are left with the body fossils, trace fossils, and our current understanding of physiology to understand the sauropod.

Paleontology gives us the bone structure of the sauropod dinosaur and insight into their behavior, based on tracks and tail drag imprints, but what was it about the inside of the sauropods that allowed them to thrive and function as such a unique species. Indeed, paleontologists have worked hard for over a hundred years trying to add context to the biology of sauropods, and they have provided evidence for essential components to their lives. I do a disservice here summarizing their main findings. There are many critical nuances to bone structure and sauropod evolution from one group, at one instance in the geological record, to the next. It is a fascinating study that I highly recommend any reader to pursue. I would point the readers to several outstanding books on this topic, including (but not limited to) *Biology of Sauropods* and *The Sauropod Dinosaur* by Mark Hallett and Matthew Wedel. Alas, these books must take a summation approach to provide the background of the field of paleontology in order to focus on the physiology.

The goal in this chapter is to provide an overall summary, a bird's-eye view, of what is known about sauropods and some of the strong pieces of evidence that have already been described. Between this chapter and Chapter 2, the known data, or strong evidence, provide an important backdrop to begin to understand the extreme physiology of sauropod dinosaurs. Further chapters try to fill in the holes with physiology and perhaps try to answer the question about how these animals became dominant over the earth for so long.

Beyond paleontology, recent advances in paleontological techniques have revolutionized our understanding of dinosaurs, sauropods in particular. Paleohistology, in particular, has played a pivotal role in deciphering the growth patterns and longevity of these ancient behemoths. Paleohistology is a specialized field within paleontology that involves the microscopic study of fossilized bone tissue. It allows researchers to gain insights into the growth, development, and physiology of extinct organisms by examining the microscopic structure of their bones.

The process of paleohistology begins with the collection of fossilized bone specimens from excavation sites. These specimens are carefully cleaned and prepared in the laboratory to remove any debris or sediment that may obscure the bone tissue. Once prepared, thin sections of the bone are cut using specialized saws or grinding equipment; these slices are then observed under a polarizing light microscope or a scanning electron microscope, for paleohistologists to study the microscopic structure of the bone tissue in detail.

This includes features such as growth rings (also known as growth lines or lines of arrested growth). Growth rings in bone tissue, similar to tree rings, can provide valuable information about the age of an individual and its growth history. By counting the number of growth rings and measuring the spacing between them, researchers can estimate the age of the organism at the time of death and infer its growth rates during different stages of life. In addition to age estimation, paleohistology can also provide insights into other aspects of an organism's biology, such as metabolism, reproductive strategies, and disease. For example, the presence of certain bone features, such as vascular canals or evidence of bone remodeling, can indicate periods of rapid growth, metabolic activity, or injury and healing.

Last, we will be talking about sauropods in aggregate, also called a clad. A clad is a term used to indicate a grouping of species based on a common evolutionary ancestor. It's an easy term to utilize when discussing sauropods because of the wide variety of sauropod-related species, but also in ease of use because generally they share a lot of similar commonalities, and it is easier to refer and discuss them in this way. We are in many ways forced to describe the sauropod animals as one species; otherwise, it becomes too unwieldly to understand the nuances of every single species. Thus, a lot of the concepts and descriptions are primarily for a 30,000 ft. view of how the dinosaur functioned. However, it should be well understood that sauropods consist of a minimum of 130 different species, with more discovered every year.

Ologies

With paleontology, we have been able to build out, literally and figuratively, the skeleton of a sauropod. The job of physiology is to fill in the skeleton to understand how the whole animal functioned. Physiology is the study of how animals maintain homeostasis—the ability for an animal to regulate independently it's body, ranging from organ function to neuronal responses, in a state of equilibrium. In other words, physiology attempts to understand the functioning of an efficient organic machine. What is the use of trying to study the physiology of an organism from greater than 65 million years ago if all we are left with are the fossilized bones? A key reason for doing this is that by trying to understand the physiology of an organism that lived in such a uniquely different environment from ours, and was so "successful" in terms of its distribution, we learn more about our own mammalian physiology, as well as physiology of more unique organisms living in the present.

One reason we can start to make inferences about sauropods' physiology based on what we currently know is that the cellular processes that underlie the tissue-specific physiological responses are likely similar to the ones our body currently utilizes. Many of the functional components across animals, extinct or not, are highly conserved. The major reason for this is likely the evolutionary drive for life processes dictated by the environment on earth itself (more on this in the next chapter). For example, this has been demonstrated using proteins called caveola. Caveolae have been well-studied in mammalian cells where they provide important atomic-scale scaffold-like structures within a cell to organize multiple proteins in specific areas.

Recent work has demonstrated that caveolae, isolated from protozoan separated by mammals by almost 1 billion years, retain this same structure and function just like the human caveola when put back into isolated human cells! This fascinating piece of data, although it's only been shown with one very unique protein, provides a wide room to allow us to extrapolate what we know about physiology now, to 65 million years ago. It also provides a framework to think about what modifications sauropods may have had on proteins to make them work more efficiently.

What about sauropods in terms of extreme physiology? Extreme physiology examines the extreme edges of homeostasis, situated between physiology and pathology. In this zone, the animal does maintain a constant homeostatic function, but it does so with adaptions that most other animals could not survive in.

In many cases, the adaptions are just enough to ensure survival, so that if any one thing is off, the animal has a difficult time adjusting to a more normal physiological environment. Examples of this include understanding how animals live and thrive in low-oxygen environments (e.g., the Himalayas), how animals perform massive aerobic feats (e.g., cheetahs running to catch a gazelle), and how animals thrive in extreme temperatures—hot (e.g., camels in the desert) or cold (e.g., the polar bears in the Artic). Sauropods are an excellent case of extreme physiology in our current epoch, helping us understand hypoxia and gigantism and how it might even apply to us.

How is a sauropod the same or different?

Ornithischia, Theropoda, and Sauropodomorpha are three major groups of dinosaurs that dominated the Mesozoic era. Each group exhibits distinctive anatomical features and ecological adaptations, reflecting their diverse evolutionary paths.

Ornithischia, characterized by their bird-like hips, encompasses a wide range of herbivorous dinosaurs. One of the key features distinguishing ornithischians is the backward-pointing pubis bone of the pelvis, which allowed their gut to process fibrous plant materials. Examples of ornithischians include the iconic *Triceratops* with its elaborate frill and horns, the heavily armored *Ankylosaurus*, and the duck-billed hadrosaurs like *Parasaurolophus*.

Theropoda, on the other hand, represent a diverse group of carnivorous dinosaurs, ranging from small, agile predators to large carnivores. Their hips exhibit a more lizard-like structure, with the pubis bone pointing forward, allowing for greater agility and speed. Theropods are characterized by their sharp, serrated teeth, powerful jaws, and clawed hands, used for grasping and tearing prey. Among the most famous members of this group are the *Tyrannosaurus* rex, the *Velociraptor*, and the bird-like *Archaeopteryx*, often considered a crucial link between dinosaurs and birds. Theropods dominated terrestrial ecosystems as apex predators, showcasing adaptations for hunting.

The sauropods had hips that resemble those of lizards, with the pubis pointing forward. One of the defining features of sauropods is their quadrupedal stance, supported by pillar-like legs and a robust skeletal structure adapted for supporting immense body weight (Fig. 1.2). There is a current controversy about how to organize these three groups with new work claiming other factors need to be considered besides the traditional hips. This includes skull size and bones of the legs. Still, when being compared with other dinosaurs, sauropods constitute a distinct group.

Sauropods, lighter than air

The sauropods, being as big as buses or five-story buildings, would not be possible if their bones were dense and heavy like ours. The weight of their bones would restrict quick or much movement at all and require a huge amount of muscle and energy to move them. Thus, like present-day birds, dinosaurs had hollow bones with inner structures known as air sacs, which

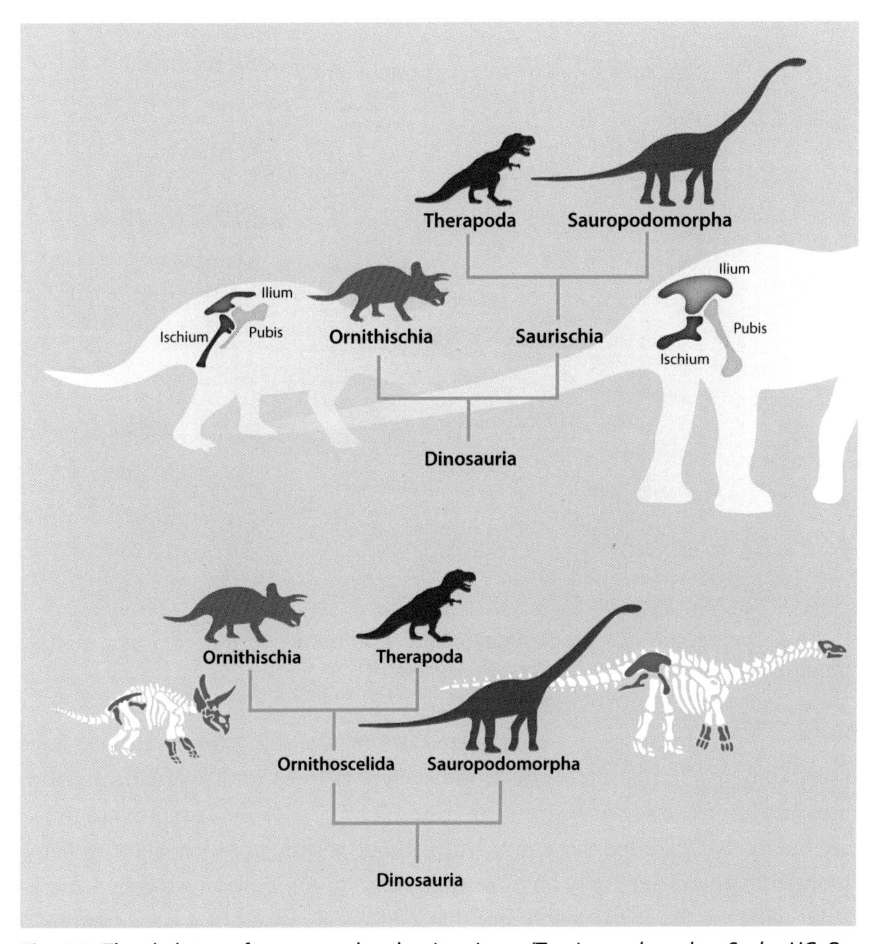

Fig. 1.2 The skeleton of a sauropod makes it unique. *(Top image based on Seeley HG: On the classification of the fossil animals commonly named Dinosauria, Proceedings of the Royal Society of London 43, 165–171, 1888, https://doi.org/10.1098/rspl.1887.0117. Bottom taxonomy proposed by Baron MG, Norman DB, Barrett PM: A new hypothesis of dinosaur relationships and early dinosaur evolution, Nature 543:501–506, 2017.)*

made their skeletons lighter and less dense. These are termed pneumatized bones, and they are found throughout the entire sauropod skeleton, providing a lightweight, but still strong, structure for tendons to hold on to (Fig. 1.3). The pneumatized bones were arguably the most essential part of a sauropod. These types of bones were apparently so advantageous that they emerged at least three times during the evolution of dinosaurs and pterosaurs (flying reptiles).

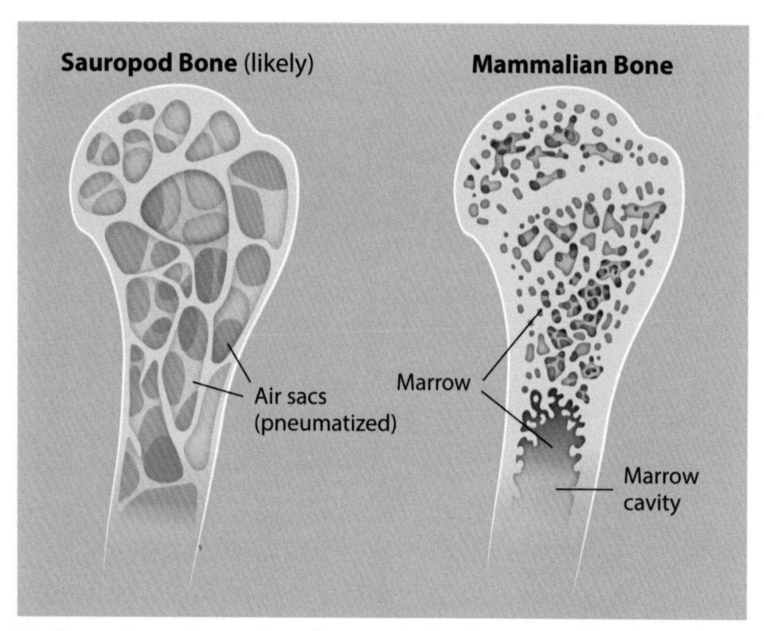

Fig. 1.3 Comparison of a modern-day mammalian bone and that of a pneumatized sauropod bone.

Another advantage of pneumatized bones was that they likely provided more oxygen circulating in the blood which allowed for a constant metabolism and overall oxygenation of the tissue. The large bone spaces were full of air, but likely facilitated more rapid movement of large volumes of oxygenated blood to tissue. There may have also been a component to breathing, remarkably, with the pneumatized bones that will be discussed in a later chapter.

Another strong possibility was that pneumatized bones played a role in regulating body temperature. This possibility may be because of two reasons. The first is the sauropod would simply require less energy to move its body and thus produce less heat. The other reason is heat and cold could be more quickly dissipated within the air spaces in the bones allowing for a large distribution of the heat.

Interestingly, the pneumatized bones were a reason many paleontologists initially thought sauropods were mainly in the water, as the trapped air in the bones are generally used by animals to float, similar to a boat. The problem with this hypothesis, in general, is the sauropods don't have a ballast that would allow them to balance in the water. For this reason, the sauropods would constantly be tipping over until there was enough water to make them float (Fig. 1.4).

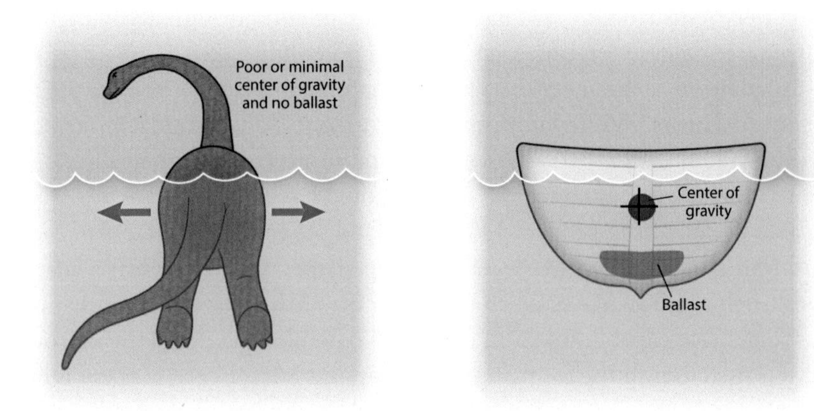

Fig. 1.4 The difficulty in balancing of a sauropod in water.

Last, the pneumatized bones create problems when recovering sauropod fossils. For example, the *Ornithopsis* species of sauropod was thought to be a pterosaur when first discovered in the 1800s due to the extensive pneumatization in the bones. This was also an issue with the discovery of *Nigersaurus taqueti*, as the pneumatized bones were poorly fossilized, making characterization difficult.

Heads or tails...or neck and tails

Setting the sauropod apart from others was its tail and long neck that support its small head. Starting at the tail, we broadly assume it was muscular to support the hind limbs, but also to stabilize the overall body size of the sauropod. Other key unknowns include how they used their tails. Did they use their tails to communicate between other sauropods, to attract mates, or maybe to fight?

The sauropod tail stretched to impressive lengths, sometimes comprising more than half of their total body length; sauropod tails were formidable structures. The tails were composed of a series of elongated vertebrae, often fused together into a rigid structure. One of the primary roles of the sauropod tail was balance. As sauropods lumbered across the Mesozoic landscapes, their long necks and equally lengthy tails acted as counterbalances, helping stabilize their bodies and prevent them from toppling over. This balance was especially critical for sauropods, given their immense size and weight. By swaying their tails in concert with their movements, sauropods could effectively redistribute their mass and maintain equilibrium, allowing them to navigate diverse terrains with relative ease.

In addition to balance, the sauropod tail likely played a role in defense against predators. While sauropods were undoubtedly formidable creatures in their own right, they were not immune to predation, especially during their vulnerable juvenile stages. The tail could have been used as a weapon, capable of delivering powerful strikes to fend off would-be attackers. Some sauropod species may have even developed specialized defensive adaptations, such as bony spikes or clubs on the tip of the tail, further enhancing their defensive capabilities. Some scientists have also suggested the sauropod tail was used in fighting for mates. The last point is less likely, as there is little evidence in the fossil record of sauropods tail breaks; however, it still can't be discounted as a defense mechanism.

Furthermore, the sauropod tail may have served as a means of communication within social groups. While the specifics of sauropod social behavior remain highly speculative, it's plausible that these animals utilized tail movements and gestures to convey information to one another, such as warnings of danger, mating displays, or territorial disputes. The ability to communicate effectively would have been essential for maintaining social cohesion within sauropod herds, particularly during migrations or other group activities. This is mostly inferred based on other species, but the length and prominence of the tail make this a real possibility.

Finally, recent research has suggested that the sauropod tail may have played a role in thermoregulation. By adjusting the position of their tails relative to the sun, sauropods could potentially regulate their body temperature, either by maximizing or minimizing exposure to solar radiation. This hypothesis is supported by anatomical evidence, indicating that sauropod tails were highly vascularized, meaning they contained numerous blood vessels that could have helped dissipate or retain heat as needed. This type of thermoregulation is also seen in animals today, lending strong evidence to this potential secondary role. Thus, the sauropod tail likely served as a multifunctional appendage, fulfilling a variety of critical roles in their lives.

Comparable to the tail is the sauropod neck. Sauropod necks have been subject to extensive scientific study to understand their biomechanics, physiology, and evolutionary significance. These necks, often exceeding lengths of 30 ft. in some species, were remarkable adaptations to reach vegetations at varying heights and distances. Composed of numerous elongated cervical vertebrae, sauropod necks displayed a wide range of morphologies, from relatively short and robust in some species to incredibly elongated and slender in others. The cervical vertebrae of sauropods were characterized by distinctive pneumaticity, with hollow spaces within the bones that reduced weight while maintaining structural strength. This pneumaticity likely contributed to the overall lightweight nature of sauropod necks, allowing them to be maneuvered with relative ease despite their immense size.

The flexibility and range of motion of sauropod necks have been topics of considerable debate among paleontologists. Studies utilizing biomechanical modeling, skeletal morphology analysis, and comparisons with extant animals have provided insights into how sauropods may have utilized their necks for feeding. It's believed that sauropod necks were capable of a wide range of motion, including vertical and horizontal movements, as well as lateral bending. This versatility allowed sauropods to exploit a diverse range of vegetation types, from ground-level ferns to tall conifers, without the need for extensive body movement. Additionally, sauropod necks likely played a role in intra-specific communication and social behavior, with different species possibly exhibiting unique neck postures and gestures for signaling and display. This too is based on what we know from giraffes, who many times lock necks for interspecies communication. It is not known if this translates to sauropods as well.

On top of the stereotypical sauropod neck is its comparatively small skull. Shapes of sauropod skull varied widely across species, dependent largely on their environment, similar to the variability in teeth (more on that in what follows). However, an interesting commonality among all sauropod species' skulls is how large the eye sockets were.

In general, eye size does scale with body size. Conversely, the blue whale has eyes that are surprisingly smaller than its body size—although the eyes are the size of grapefruits, they are still small for its overall size. Elephants also have rather small eyes for their size, but they are exceptionally sensitive to blue and violet wavelengths of light, allowing them to travel at night. In present day, large eyes are strongly associated with nocturnal behaviors of the animal. This is because the larger eyes allow for more light to be gathered, so that even under low-light conditions, even a few number of photons would be observable by the animal. Nocturnal animals are generally smaller and strongly trend toward being a predator or prey.

Sauropod's size alone, both in terms of not being able to hide at night and not being a reliable meal for a carnivorous dinosaur, is evidence against nocturnal behavior. There is also no evidence in the fossil record to suggest sauropods were nocturnal. For such an enormous animal, the fossil record would suggest that eyesight was an important component to their day-to-day life. The giraffe, with its long neck, has exceptionally large eyes. Horses share this physical attribute of large eyes perched on a skull protruding off a long neck. Horses and giraffes share a lot of similarities with their eyesight, including a large range of vision. Both species can see completely around themselves (i.e., almost 360 degrees), with only the absolute direct front and back of the animals being a blind spot for them. Both species can also use their eyes together to focus on objects at a distance, like binoculars. One of the most peculiar things about horse and giraffe eyes, however, is that each eye can work independent of each other, and the item being viewed in one eye can be processed differently than the item in the other eye. One can imagine this type of sight on such a large animal as a sauropod would be quite advantageous.

All that being said, it isn't clear how strong the eye muscles were in sauropods for their relatively large size. It is possible their eyes may have been large, but a small skull may have prevented development of many muscles to control the eyes or provide space for the muscles that were there to control the eyes quickly. This could imply a delayed ability to focus for sauropods because of weak eye muscles, which means they couldn't focus on objects quickly.

Another notable feature of the sauropod skull is the arrangement of their nasal openings. Sauropods had large external nares located near the top of their skulls, suggesting that they had an extensive nasal cavity. This arrangement may have allowed for efficient cooling of the brain, as well as providing a keen sense of smell to help locate food sources over long distances.

Taking a bite out of the sauropod

One of the best fossils we have of sauropods that provides direct insight into their physiology are their teeth (as Lhuhyd had found on the beach). This makes sense in terms of their dense structure being ideal for fossilization and the number of teeth sauropods went through throughout their lifetime.

Sauropods had an incredible array of teeth, each with their own shape and purpose. Sauropod teeth exhibited a unique dental arrangement called heterodonty, meaning they had different tooth shapes within their jaws. Toward the front of the mouth, they had spatulate teeth, which were broad and spoon-shaped. As you move toward the back, the teeth transitioned into cylindrical or peg-like shapes. With these "peg-like" teeth, sauropods could easily maneuver their mouths to pluck leaves from trees or shrubs.

This dental diversity allowed sauropods to efficiently process various types of vegetation, from soft leaves to tougher plant parts like stems and branches.

The enamel covering sauropod teeth was remarkably thick compared with other dinosaurs. It acted as a protective layer, shielding the underlying dentin and pulp from wear and tear. The enamel was composed of hydroxy-apatite crystals, which provided strength and resistance to mechanical stress. This robust enamel allowed sauropods to repeatedly bite and chew tough plant material without damaging their teeth.

Sauropod teeth were continuously replaced throughout their lives, a process known as polyphyodonty. The replacement teeth grew within the jawbone and gradually pushed forward, eventually replacing the worn-out teeth. This constant tooth replacement ensured that sauropods always had functional teeth available for feeding, even if some were lost or worn down. It also contributed to the longevity of their dental batteries, allowing them to sustain their herbivorous diet over long periods.

The dental batteries of sauropods were also impressive structures. The teeth were tightly packed together in rows, forming a continuous cutting edge. This arrangement increased the efficiency of food processing, as the teeth worked together to slice through vegetation. The dental batteries were also vertically stacked, allowing sauropods to effectively shear plant material as they closed their jaws, similar to the action of scissors.

Microscopic studies of sauropod teeth have revealed distinct wear patterns. The wear facets on the teeth suggest that sauropods employed a grinding motion while feeding. As they bit down and moved their jaws from side to side, the teeth ground against each other, breaking down plant material into smaller particles. This grinding action, combined with the dental battery's continuous replacement, enabled sauropods to extract maximum nutrition from their plant-based diet. By having these specialized teeth, sauropods could strip leaves off branches or twigs, choosing the specific parts of plants they preferred to eat. This ability to selectively feed on certain vegetation might have allowed sauropods to obtain the nutrients they needed while avoiding less nutritious or less desirable parts of plants.

Sauropod short ribs

In many ways, the spine of the sauropod could be imagined as a superhighway across its large size. The presence of complex vascular canals within the vertebrae, independent of the air sacs, suggests a highly developed blood supply, essential for delivering oxygen and nutrients to the immense tissues of these giants. The arrangement of the cervical and dorsal vertebrae allowed for a fluid, wave-like motion as the sauropod walked. Despite their immense size, sauropods were remarkably agile animals, thanks in part to the structure of their vertebrae facilitating efficient locomotion. Although sauropods are often depicted as slow-moving, lumbering creatures, recent studies suggest that they were capable of surprising speed when necessary.

One important question in terms of the boney structure of sauropods is how to support the weight of the neck. With the defining characteristic of the sauropods across all the different species being the long neck, there had to be a way to support that essential anatomical structure whether the neck was raised, held lateral to the ground, or even if bent to the ground. How could they support the weight to raise it consistently, for instance, for feeding, pulling off leaves, or looking for predators? Muscles alone wouldn't necessarily be the best way to do this for a few reasons: one is the excessive weight of the muscles needed to hold the neck in place which would make holding the neck even more problematic. Practically, this would also require a much higher-energy expenditure because of the high metabolic load of load-bearing muscles. There are a few different solutions to the neck-raising or neck-lowering problem that sauropods utilized, which certainly included pneumatized bones, as discussed before, to lighten the weight, but specific to spinal vertebrate, this also likely included a bifurcated spine.

Bifurcated spines were found throughout sauropods and formed a prominent ridge along the back. The spines consisted of long bony protrusions coming off the vertebrate that created a "U"- or "V"-like shape on the posterior side of the spine. Conceptually, this bifurcation would allow ligaments to attach on either side and help support the neck weight. But not all the spines were bifurcated, only in certain spots along the spine, usually in the thoracic region. It's interesting these bifurcated spines on the vertebrate are a common evolutionary solution to support heavy necks, including those seen in ox and cows with extra heavy horns and skulls.

Along the spine, sauropod ribs were integral to supporting the colossal bodies of these giant dinosaurs and played a crucial role in their unique physiology. One of the most striking features of sauropod ribs is their immense size, which matched the scale of their bodies. These ribs were elongated, curved bones that formed a framework around the chest and abdominal cavity, providing structural support and protecting vital organs. Their size and shape allowed for the expansion of the ribcage, accommodating the vast volumes of air required to fuel the sauropods' high metabolic rates and sustain their massive bodies.

The arrangement and curvature of sauropod ribs also reflect their unique biomechanics and posture. The ribs of sauropods were typically broad and flat, curving downward and outward from the vertebral column. This configuration provided ample surface area for the attachment of muscles involved in respiration and locomotion, allowing sauropods to support their massive bodies while moving efficiently. Additionally, the curvature of the ribs likely contributed to the barrel-shaped chest characteristic of sauropods, further enhancing their much needed high respiratory capacity.

Growing-up sauropod

Sauropods did not have live births; they laid eggs. While rare in the fossil record compared with those of smaller dinosaurs, sauropod eggs offer valuable insights into the reproductive biology and life history. Discovered eggs typically range in size from football to basketball size, with some species likely laying eggs even larger than that. The shape of sauropod eggs varies, but they generally exhibit an elongated or elliptical form, resembling the eggs of modern reptiles. These eggs were likely laid in nests, constructed by sauropod parents in suitable environments such as riverbanks or floodplains. The construction of these nests may have involved scraping out depressions in the substrate and layering them with vegetation for insulation and protection. Some fossilized sauropod nesting sites provide evidence of communal nesting behavior, suggesting that these dinosaurs may have gathered in groups to lay their eggs, possibly for added protection against predators.

The study of sauropod eggs has revealed interesting details about their reproductive strategies and parental care. Analysis of eggshells has shown evidence of microstructures indicative of rapid egg formation, suggesting that sauropods laid eggs in large clutches, possibly in quick succession. This reproductive strategy, coupled with communal nesting behaviors, may have been adaptations to maximize offspring survival in the face of high predation pressure. Additionally, the thickness and structure of sauropod eggshells suggest adaptations for protecting the developing embryos from environmental fluctuations and physical damage.

There is a great deal of debate about sauropod lifespans, mainly having to do with whether a sauropod is classified as a juvenile or an adult. As one can imagine, this can create confusion for assessing lifespans for these animals. It is generally agreed sauropods grew up very quickly from hatchling, a concept discussed further in the next chapter. The lifespan of sauropods was likely influenced by a myriad of factors, including environmental conditions, resource availability, and reproductive strategies. Their enormous size may have conferred certain advantages, such as reduced predation risk from smaller predators, but it also presented challenges in terms of resource acquisition and energy expenditure. Evidence suggests that sauropods could migrate over vast distances in search of food, which may have contributed to their longevity by allowing them to exploit a wide range of habitats. Additionally, the reproductive strategies of sauropods, such as nesting behaviors and parental care, would have influenced the survival rates of offspring and, consequently, the overall population of the animals. In general, it's now thought sauropods lived commonly around 50 years, give or take a few decades.

Giant shoes to fill

It is now abundantly clear sauropods were huge animals with several specialized adaptions in their body composition and overall structure to accommodate their gigantism. A part of their body we have yet to discuss are their legs, supporting these giant structures. How did the sauropod walk with this type of weight and not crack its femurs with every step? In a recent study of several different sauropod species, it was found that regardless of their skeletal leg posture, all examined specimens would have struggled to support their weight without a soft tissue pad in the foot—akin to the cushion in a sneaker. The absence of this pad resulted in maximum bone stresses exceeding critical levels, suggesting a high probability of mechanical failure, such as bone stress fractures, especially under repetitive heavy loads during normal, day-to-day locomotion. Conversely, the presence of a soft tissue pad at the base of the sauropod foot would have significantly reduced bone stress, indicating it was likely these would have played a critical role in absorbing and releasing energy, potentially aiding in the sauropods' success as a species. In fact, despite different postural configurations among all the different species of sauropods, subtle similarities exist in bone stress patterns likely existed among them. This uniformity in stress patterns across various postures points toward the possibility that sauropods may have utilized other extrinsic soft tissues, along with distinct weight distribution mechanisms, to further mitigate stresses on their foot bones (Fig. 1.5).

From an evolutionary perspective, it is likely the acquisition of the soft tissue pad at the bottom of the foot likely occurred early in sauropod evolution, allowing for a functionally plantigrade foot structure while retaining

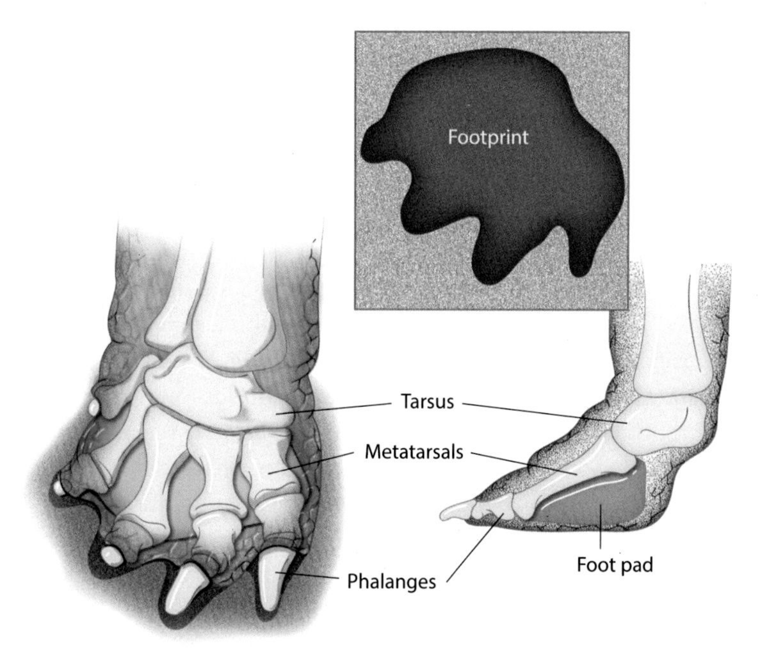

Fig. 1.5 The possible foot of a sauropod. *(Image based on Jannel A, Salisbury SW, Panagiotopoulou O: Softening the steps to gigantism in sauropod dinosaurs through the evolution of a pedal pad, Sci Adv 8:eabm8280, 2022.)*

diverse skeletal pedal postures. This adaptation potentially facilitated their ability to achieve immense body sizes by the Late Triassic-Early Jurassic period. In addition, the soft tissue under the foot of sauropods provides further evidence the animals were likely land-dwelling because this type of reinforcement wouldn't have been necessary in water dwelling animals.

Connections to sauropod physiology have also begun to be put together with their footprints. Trace fossils of sauropod trackways and other individual footprints, known as "ichnites," are widely distributed across various continents and locations, giving us ample evidence that sauropods could be found almost everywhere geographically. They also suggest multiple behaviors that influence their physiology. For instance, sauropod tracks discovered in the Villar del Arzobispo Formation in Spain suggest gregarious and social behavior among these dinosaurs. More importantly, these trace fossils offer evidence supporting hypotheses regarding sauropod fore and hind foot anatomy. Typically, forefoot prints are significantly smaller and often crescent-shaped compared with hind footprints. Occasionally, ichnites even preserve traces of claws, aiding in the identification of sauropod groups that lost claws or digits on their forefeet. However, determining whether the footprints belong to juveniles or adults remains challenging due to the lack of previous trackway individual age identification.

Sauropod trackways are typically categorized into narrow, medium, and wide gauges based on the distance between opposite limbs, which are indications of limb configuration and locomotion. Research indicates that advanced sauropod groups exhibit distinctive trackway gauges, with each sauropod family characterized by specific patterns. Narrow-gauge limbs, often featuring strong impressions of large thumb claws on the forefeet, are common among most sauropods except titanosaurs. Medium-gauge trackways with forefoot claw impressions likely belong to brachiosaurids and other primitive titanosauriformes, while primitive true titanosaurs also retain forefoot claws but exhibit wider-set limbs. Advanced titanosaurs, on the other hand, typically have wide-gauge limbs devoid of claws or digits on the forefeet.

In some cases, only trackways from the forefeet are preserved. Initially, this was thought it was because the back of the sauropod was floating up in water. However, as mentioned before, with the lack of evidence of sauropods submerged in large amounts of water, this may be a limited occurrence. More recent computer modeling suggests that the preservation of tracks may depend on specific substrate conditions, with differences in hind limb and forelimb surface area influencing contact pressure and preservation outcomes.

Overall, the trace and body fossils of sauropods have provided a large amount of insight into their possible physiology, but many questions remain unanswered. We will start to connect the dots by examining the environment in which sauropods lived as a first step in putting together the rest of their possible physiology.

References and further reading

Baron MG, Norman DB, Barrett PM: A new hypothesis of dinosaur relationships and early dinosaur evolution, *Nature* 543:501–506, 2017.

Benoit J, Penn-Clarke C, Helm C: Africans discovered dinosaur fossils long before the term 'paleontology' existed, *The Conversation* (January 4), 2024.

Curry Rogers KA, Wilson JA, editors: *The sauropods*, Berkley and Los Angeles, California, 2005, University of California Press.

Hallett M, Wedel MJ: *The sauropod dinosaurs*, Baltimore, Maryland, 2016, John Hopkins University Press.

Jannel A, Salisbury SW, Panagiotopoulou O: Softening the steps to gigantism in sauropod dinosaurs through the evolution of a pedal pad, *Sci Adv* 8:eabm8280, 2022.

Klein N, Remes K, Gee CT, Sander PM, editors: *Biology of the sauropod dinosaurs*, Indianapolis, Indiana, 2011, Indiana University Press.

Woodruff DC: The anatomy of the bifurcated neural spine and it's occurrence within tetrapoda, *J Morphol* 275:1053–1065, 2014.

CHAPTER TWO

The environment of the sauropod and its physiology

We will continue from the previous chapter, which focused on what we know about the fossil record and the broad conclusions we can draw about the biology and possible physiology of sauropods, and attempt to add a little bit more in this chapter with evidence from the environment the sauropods lived in. This is because the environment strongly shapes the evolutionary history and adaptions species obtain in order to be in homeostasis with their environment.

The concept of homeostasis is a fundamental component of physiology. It's how the body is able to work in balance among all the organs and the environment the species lives in. There are four essential factors of homeostasis in all multicellular, nonplant organisms. They include (1) consistent blood volume, (2) fluid volume among the organs, (3) metabolism (body temperature), and (4) the animals' tissue ph. To put this into context, blood volume is essential for the delivery of oxygen to tissues, to ensure consumed food can be utilized for energy by a consistent metabolism. The metabolism as well as other essential chemical reactions in the body is tightly controlled by the pH, whose maintenance ensures the enzymes work at a consistent rate. Fluid volume helps regulate body temperature and clears by-products of metabolism. The four components of homeostasis are controlled both genetically (intrinsic to the organismal species) and by the environment the organisms live in.

The environment an animal lives in greatly influences their physiology in short and long evolutionary time scales. This is evident in humans living at high altitudes where oxygen concentration is low (hypoxia, or low oxygen, whereas at ground level, oxygen concentration in the atmosphere is ~22%). Humans living at high altitudes, such as those in the Andes, Himalayas, and Ethiopian Highlands, exhibit several physiological adaptations to cope with lower oxygen levels. These adaptations include increased red blood cell production and higher hemoglobin concentrations, which enhance the blood's oxygen-carrying capacity (see further chapters). Additionally, high-altitude dwellers often have larger lung volumes and more efficient breathing patterns, to maximize oxygen intake. Their bodies also exhibit increased capillary

Balancing a Sauropod
https://doi.org/10.1016/B978-0-12-823303-0.00003-3

density in muscles, improving oxygen delivery to tissues. Moreover, there are genetic adaptations, such as variations and mutations in genes, that have been measured among Tibetans, which help regulate the body's response to hypoxia. These changes collectively enable individuals to maintain adequate oxygenation and physical performance in environments where oxygen is scarce. This has generally happened in a time scale on the order ~10,000 years, likely at the longest. On the geological time scale, this is miniscule.

Butterflies are believed to have evolved from moths, through a series of gradual adaptations driven by ecological and environmental pressures. The evolution likely began during the Late Cretaceous period, around 100 million years ago, when flowering plants (angiosperms) started to diversify significantly. Moths, which were primarily nocturnal, began to exploit new ecological niches by transitioning to diurnal (daytime) activity. This shift likely offered several advantages, including access to new food resources, such as the nectar of flowering plants, and reduced predation pressures at night. Over time, these diurnal moths developed brighter colors, which served both as a means of attracting mates and as a form of camouflage or warning to predators. Additionally, the structure of their wings evolved to become more efficient for daytime flight, often becoming broader and more colorful to enhance thermoregulation and visual signaling. These changes, coupled with the development of more specialized feeding structures like elongated proboscises for sipping nectar, gradually gave rise to the distinct lineage we now recognize as butterflies. Genetic mutations and natural selection played crucial roles in this evolutionary process, leading to the diverse and widespread group of insects we see today.

It should be noted that we are referring to two different things, adaptation and evolution. Adaptations happen in short time scales, and evolution happens in long, million year time scales in many cases due to adaptations. Adaptations are traits or characteristics that enhance an organism's ability to survive and reproduce in a specific environment, for example, humans living in hypoxic conditions with more red blood cells to carry oxygen. These traits can be structural, physiological, or behavioral and arise from genetic variations within a population. Evolution, on the other hand, is the process through which populations of organisms change over time due to changes in allele frequencies, or the types of genes, within the genetic makeup of entire groups of animals. An example of this is how the genetic composition of butterflies became unique from moths due to multiple adaptions to exploit the environment the moth/butterfly was living in.

Here, we focus on the environment the sauropods lived in and start to understand how it may have prompted unique adaptions to alter their physiology and make them the sauropod species that were so successful around the world. To do that, we need to get a clear idea of the time and place of when the sauropods lived.

Back in time

Humans attempt to control their physical environment all the time. We build and destroy, to make our lives more sustainable and even more comfortable. Of course, this wasn't an option for sauropods, which left them completely dependent on the physical environment to live. And their ability to survive and find homeostasis was dependent on the environment which dynamically changed during the time period they lived in. Thus, to understand how the physiology of sauropods could have existed, we must start with a background of the environment these beasts lived in for 150 million years and possibly the environmental changes as their species radiated as well as contracted.

The Mesozoic Era on earth, spanning from ~252 to 66 million years ago, was marked by significant geological, climatic, and evolutionary changes, characterized by distinct environmental conditions across its three major periods, including the Triassic, Jurassic, and Cretaceous.

1. Triassic (252–201 million years ago): During the early Triassic period, the earth was still recovering from the mass extinction event at the end of the Permian. The climate was generally arid and hot, with vast desert regions dominating the landscape. Large inland seas were present, particularly in low-lying areas, fostering diverse marine ecosystems. On land, the flora consisted mainly of ferns, conifers, and early gymnosperms, which gradually diversified as the Triassic progressed. Presauropod-like species were just evolving at this point.
2. Jurassic (200–145 million years ago): Moving into the Jurassic period, the climate became warmer and more humid, compared with the preceding Triassic. This period is often associated with lush tropical forests, extensive shallow seas, and abundant marine life. Pangea, the supercontinent, began to break apart, leading to the formation of separate landmasses and distinct ocean basins. Dinosaurs, including iconic species like *Stegosaurus* and *Allosaurus*, thrived in this environment, alongside early mammals, reptiles, and flying reptiles like pterosaurs. Sauropods reached their

zenith, in terms of a huge range of different species in the Jurassic period, and remained stable but slowly drop off after that.

3. Cretaceous (145–66 million years ago): The Cretaceous period witnessed further fragmentation of landmasses and the emergence of modern continents. The climate was generally warmer and more stable than in previous periods, with polar ice caps much smaller or absent. The interior of continents experienced seasonal variations, while coastal regions were characterized by humid conditions and extensive deltas. The global sea level was higher than today, resulting in widespread shallow seas and epicontinental waterways teeming with diverse marine life, including ammonites, marine reptiles like ichthyosaurs and plesiosaurs, and early fish species. Sauropods species were suspiciously far fewer in this time period, having witnessed a large extinction event at the end of the Jurassic. Why they never recovered is not known (Fig. 2.1).

Throughout the Mesozoic Era, volcanic activity was a significant geological process, contributing to the shaping of landscapes and influencing climate

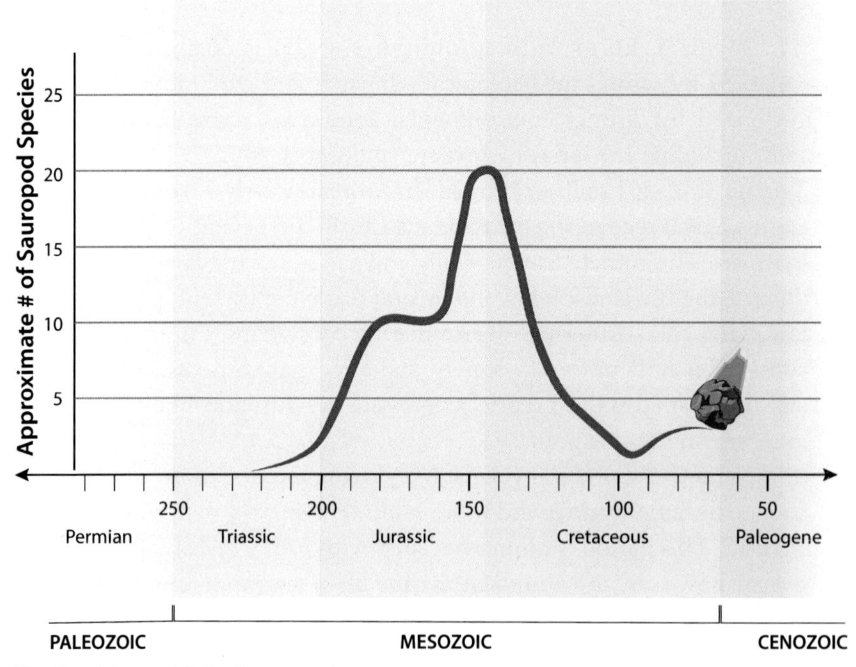

Fig. 2.1 Rise and fall of sauropods.

patterns. Additionally, tectonic activity, such as the breakup of Pangea and the formation of new mountain ranges, played a crucial role in shaping terrestrial and marine environments.

What's shaking?...Everything

During the Mesozoic Era, tectonic activity played a pivotal role in shaping the earth's surface, leading to the breakup of the supercontinent Pangaea and the formation of modern continents and ocean basins. The era began with the early Triassic, characterized by the aftermath of the Permian mass extinction event and continued with significant tectonic activity. One of the prominent events was the rifting of Pangaea, initiated during the Triassic and culminating in its eventual breakup during the Jurassic and Cretaceous periods. This fragmentation led to the formation of separate landmasses, including Laurasia in the north and Gondwana in the south, which later evolved into the continents we recognize today.

The breakup of Pangaea resulted from the movement of tectonic plates, driven by processes such as mantle convection and plate boundary interactions. As Pangaea began to rift apart, new ocean basins formed between the separating landmasses, including the Central Atlantic Ocean and the Indian Ocean. This rift also created extensive rift valleys, such as the famous East African Rift System, which continues to evolve today near Ethiopia. The separation of continents altered oceanic circulation patterns, climate dynamics, and biogeographical distributions, influencing the evolution of life on Earth.

Subduction zones, where one tectonic plate moves under another and sinks into the mantle, were active around the margins of continents, leading to the formation of more volcanic arcs and eventual formation of mountain ranges as land masses came together. These zones have significant geological activity, including the formation of deep ocean trenches, volcanic arcs, mountain ranges, and earthquakes. For example, the Andes Mountains in South America began to form during this era because of the subduction of the oceanic Nazca Plate beneath the South American Plate. Similarly, the Sierra Nevada Range in North America and the Alps in Europe experienced uplift due to tectonic processes. These mountain-building events not only reshaped landscapes but also influenced global climate patterns by affecting atmospheric circulation and ocean currents.

The birth of mountain ranges and changing land geography brought the rise of reptiles as dominant land animals, the emergence and diversification of dinosaurs, and the evolution of flowering plants (e.g., angiosperms).

Welcome to sea world

Tectonic activity and changing continental configurations not only created mountains as landmasses bumped against each other, but they also caused landmasses to separate, forming new ocean basins as well as inland seas, which created opportunities for proliferation of sea life. Both oceans and inland seas played a significant role in shaping terrestrial and marine ecosystems.

One of the most important is the Western Interior Seaway, which extended across North America during the Late Cretaceous period. This vast sea separated the continent into eastern and western landmasses, connecting the Arctic Ocean to the Gulf of Mexico. The Western Interior Seaway was characterized by shallow waters, with depths ranging from a few to several hundred meters, and featured diverse marine life, including ammonites, mosasaurs, and sharks. Sedimentary deposits from this inland sea, such as the Niobrara Formation, provided some amazing creatures and insight into the Mesozoic marine ecosystems and paleoenvironmental conditions.

Similarly, during the early to mid-Jurassic period, the Tethys Sea formed a vast inland sea stretching from the western Tethys Ocean to the Eastern Pacific. This sea separated the supercontinent Pangaea into Laurasia to the north and Gondwana to the south. The Tethys Sea was characterized by warm, tropical waters and supported a rich array of marine life, including coral reefs, ammonites, and ichthyosaurs. Sedimentary rocks deposited in the Tethys Sea, such as limestone and shale, contain abundant fossil evidence of Mesozoic marine organisms and provide valuable information about ancient oceanic conditions and biodiversity.

The new presence of inland seas during the Mesozoic Era influenced global climate patterns, ocean circulation, and the distribution of marine species, contributing to the evolutionary dynamics of life on Earth.

Changing temperatures

With rising mountain terrains, the earth experienced new diverse temperatures and oxygen level changes. In the Triassic, the climate was generally arid and hot. The Earth's continents joined together; there were extensive desert regions in the interior and monsoonal climates along the coastal areas. As the Jurassic started, the climate became more humid because of the rift valleys and shallow inland seas that had begun to form. This also led to a more varied climate with regions experiencing tropical and subtropical conditions. As the Jurassic continued, the climate continued to warm, and humidity levels increased. Extensive shallow seas covered parts of the continents, contributing to a greenhouse climate with high levels of atmospheric carbon dioxide. The early Cretaceous was also warm and humid, but as the era continued, the temperatures climbed to a warmth that would be extremely uncomfortable to modern animals (an earth average of $>25°C$ daily, year-round, compared with modern-day Earth average of $\sim 15°C$ daily, year-round).

As one can imagine from all the warmth on earth during the entire Mesozoic Era, there were no polar ice caps. Although it was warmer around the equator of the earth, it was slightly cooler at the earth poles. This data, coupled with the location of sauropod fossils, has led some to extrapolate about how sauropods regulated their body temperature, that is, whether they were endothermic or exothermic.

At its most basic, an endothermic animal produces its own heat, usually through a constant metabolism that's required for day-to-day living. The metabolic heat is derived mostly from digestion, breakdown of food, breathing, and a constant heart rate. Endothermic animals include all mammals and birds. Alternatively, exothermic is when the animal's body temperature is regulated by the external environment; also called "cold-blooded" animals. Exothermic animals include snakes and crocodiles. This is why you see many reptiles laying in the sun during the day; they do this to increase metabolism and heart rate.

There are advantages and disadvantages of being endothermic or exothermic. Exothermic animals, such as reptiles and amphibians, do not require a significant amount energy to maintain a constant body temperature. They rely on external sources of heat, such as sunlight, to reduce their metabolic demands and allow them to survive on less food or go through longer periods without needing a food source. However, exotherms' activity levels and physiological processes are heavily influenced by ambient temperatures. In cooler conditions, they become sluggish and less capable of escaping predators or catching a prey, which can limit their geographic distribution to warmer climates. Their reliance on external heat sources restricts their active periods to times when environmental temperatures are favorable, often limiting their ability to hunt, forage, or reproduce during colder periods. Endothermic animals can maintain a stable internal body temperature regardless of external conditions, allowing them to inhabit a wide range of environments, from arctic regions to tropical zones. This is due to much higher metabolic rates to sustain prolonged periods of activity, support rapid growth, and maintain high levels of physical performance, which are beneficial for predation, to escape from predators, and for long-distance migration. A disadvantage of endothermy is that maintaining a constant internal temperature requires substantial energy, necessitating a consistent and abundant food supply. This high metabolic demand can be a disadvantage in environments where food resources are scarce or unpredictable. For this reason, endotherms are more vulnerable to starvation if they cannot meet their energy requirements, as their high metabolic rates do not allow for long periods without food, unlike exotherms.

So what about sauropods, where did these beasts land in terms of body temperature? By integrating fossil evidence with climate data from the era and information on continental drift, Dr. Chiarenza et al., from the University College London, in a recent paper has inferred that sauropods inhabited warmer, drier environments compared with other dinosaurs. These habitats were probably open and semiarid, resembling modern-day savannahs. Conceptually, they also postulated that a long neck was also an advantage in heat dissipation. Their conclusion was that sauropods were more likely exothermic. Other physiological adaptions may support this conclusion. For example, some scientists claim a role for feathers in regulating body temperature. This is a key way in which birds are endothermic. There is no evidence of feathered sauropods, unlike some of the other dinosaur species. This would be a huge investment in energy for sauropods with minimal payback in terms of regulating their body temperature. In addition, one of the most fundamental ways in which mammals and birds regulate body temperature is the use of nasal turbinates. Also known as nasal conchae, these long, narrow, curled bone structures are located inside the nasal cavity. These structures are covered with mucous membranes and warm the inspired air. Thus far, there have been no reported findings of nasal turbinates in sauropods.

However, none of these observations implicitly rules out sauropod endothermy. For example, sauropods had an incredibly rapid growth rate to reach their gigantic size, requiring a high metabolic rate. In addition, there were never any issues with food sources, and the sauropods were known to migrate, some, surprisingly, based on footprints, at high speeds. There were also a few predators for sauropods based on their size. Most scientists now believe their body temperature was more on a sliding scale than an absolute endothermic or exothermic. This will be discussed in later chapters as well.

New menu changes

As the landscape and seascape changed, so did the vegetation and the food available to sauropods. During the Mesozoic Era, ferns, conifers, and early gymnosperms were prominent components of terrestrial vegetation, playing crucial roles in shaping Mesozoic ecosystems and laying the groundwork for the evolution of modern plant groups. Ferns, characterized by their vascular tissue and spore-producing structures, were among the earliest plants to colonize land and were abundant during the Mesozoic. They flourished in diverse habitats, ranging from moist, tropical forests to arid, upland environments. Ferns formed lush understories beneath towering conifers and other gymnosperms, providing habitat and food for various organisms.

Conifers, a group of seed-producing plants characterized by their needle-like or scale-like leaves and typically woody cones, were widespread and diverse during the Mesozoic Era. They dominated many terrestrial ecosystems, particularly during the Triassic and Jurassic periods, and played essential roles in stabilizing soils, regulating water cycles, and providing habitats for diverse flora and fauna. Conifers exhibited a wide range of growth forms, from towering trees like Araucaria and Sequoia to shrubby species adapted to harsh environmental conditions.

Early gymnosperms, including cycads, ginkgoes, and gnetophytes, were also prominent during the Mesozoic Era, displaying a variety of morphological and ecological adaptations. Cycads, characterized by their palm-like appearance and compound leaves, were particularly abundant and diverse during the Jurassic and Cretaceous periods. They formed dense forests in tropical and subtropical regions, providing habitat and food for a wide range of herbivorous organisms. Ginkgoes, represented today by a single living species, *Ginkgo biloba*, were more diverse during the Mesozoic and formed dense stands in temperate regions. Gnetophytes, a group with affinities to both conifers and flowering plants, were less common but nonetheless present in Mesozoic ecosystems. All of these plants were delicious to sauropods.

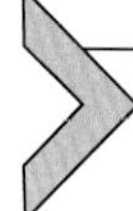

Changing habitats = changes to the menu = changes to physiology

As mentioned before, in the Mesozoic Era, the air composition was quite different than it is today. Back then, it had significantly higher levels of CO_2 and lower levels of oxygen. The higher level of carbon dioxide—estimated to be 4–6 times more than the current CO_2 levels—allowed plants to thrive and grow bigger. This led to the development of huge forest and abundant vegetation. Increased plant growth had effects on sauropods, too, with niche specialization while eating. Sauropods thrived in this era, and they had adjusted especially well to the Mesozoic ecology they lived in. As shrubs evolved taller and trees formed, this offered new food sources, if only you could reach them.

The evolutionary significance of the necks of sauropods lies in their role in shaping the ecological interactions and evolutionary trajectories of Mesozoic ecosystems. By accessing high-growing vegetation unreachable by other herbivores, sauropods occupied a unique niche as high-level browsers, influencing plant evolution and ecosystem structure. The vertebral anatomy of sauropods also reflects their unique feeding behaviors. Their neck vertebrae were highly flexible, allowing them to browse vegetation at varying heights and reach foliage that other herbivores couldn't access. This flexibility, combined with their long necks, provided sauropods with a wide feeding range, enabling them to exploit resources across diverse landscapes (Fig. 2.2).

In the same way that changes in taller plants affected dinosaurs, longer necks and taller dinosaurs affected plants. The presence of sauropods likely led to the evolution of tall, sturdy plants capable of withstanding browsing pressure from these giant herbivores. Furthermore, the sheer size and reach of sauropod necks may have provided a form of protection against predation, allowing them to feed safely out of reach of most contemporary predators. A win-win! Essentially, longer necks equated to more food and to more defensive abilities.

Longer necks weren't the only evolutionary outcome. The sauropod species *Diplodocus* exhibited elongated snouts with spoon-shaped tips, likely used for stripping leaves from branches or scooping up vegetation from the ground. Others, like *Brachiosaurus*, had shorter, more robust skulls with nasal openings positioned further back, possibly indicating different feeding strategies or ecological niches.

Separate from the huge conifers and gymnosperms, a plant that has been mentioned to have evolved due sauropods, but not eaten by sauropods, are angiosperms, which produce self-contained seeds within an ovary. We know them today as flowering plants. Definitive fossil evidence for angiosperms only dates to around 135 million years ago. Recently, however, researchers Xin Wang and Zhong-Jian Lui reported the discovery of a potential Early Jurassic angiosperm species named *Nanjinganthus*. Their work suggested a substantial similarity to ancestral models and Early

Fig. 2.2 Ecological eating niches on of a sauropod on a conifer, including the ancient angiosperm *Nanjinganthus* on the ground.

Cretaceous angiosperms, indicating that *Nanjinganthus* could indeed be a Jurassic flowering plant. The discovery of *Nanjinganthus* backed by many specimens, suggests an earlier appearance of angiosperms, potentially reshaping our understanding of their evolution and providing a further source for sauropods. Indeed, from the wastes of sauropods, there is evidence of a "grittiness," which suggests they may have been eating closer to the ground, possibly these particular ground-dwelling angiosperms (Fig. 2.2).

However, there does not appear to be any direct correlation between angiosperm emergence and significant changes in sauropod evolution, as was suggested in a seminal paper from the late 1970s. The initial hypothesis that sauropods might have played a significant role in the rise of angiosperms was based on observations of the coinciding evolutionary timelines and potential ecological interactions between the sauropods and angiosperms, in particular, the rise of angiosperms overlapping with the peak diversity and abundance of sauropods at the end of the Jurassic, beginning of the Cretaceous. Because sauropods could reach high vegetation and cover large areas in search of food, the hypothesis suggested their extensive browsing behavior might have exerted selective pressures on the vegetation.

It was suggested that sauropods could have aided in the dispersal of angiosperm seeds. As these dinosaurs moved across vast distances, they might have ingested seeds along with foliage and later excreted them in different locations, facilitating the spread of angiosperms. Additionally, the feeding activities of sauropods might have influenced the composition and structure of plant communities. Their browsing could have created open spaces and reduced competition among plants, potentially allowing faster-growing and more opportunistic plants, such as angiosperms, to proliferate. There was also the notion that angiosperms and sauropods might have engaged in co-evolutionary relationships, where the evolution of flowering plants might have been driven by the need to attract specific pollinators or to adapt to the browsing pressures exerted by large herbivores like sauropods.

However, as more detailed fossil evidence became available, the direct causal link between sauropods and the rise of angiosperms became less clear. Fossilized gut contents and coprolites (fossilized dung) provided limited evidence of direct interactions. Other factors, such as changes in climate, geology, and the emergence of other plant groups, also played significant roles in the diversification of angiosperms. The appearance of angiosperms might have been influenced more by these broader environmental changes than by interactions with sauropods alone (e.g., insects and the rise of small mammals). Modern paleobotany and paleontology suggest that the rise of angiosperms was likely driven by a complex interplay of ecological, climatic, and geological factors. While sauropods might have had some influence on the early spread and diversification of flowering plants, they are now considered one of many contributing factors rather than the primary cause. What is more likely is sauropods saw new plans such as *Nanjinganthus* as delicious snacks.

Body size

All of the food the sauropod was eating was a direct component of the environment feeding into its body size. While we will discuss metabolism and calories from these plants in later chapters, here, we will just consider, again, the enormity of the sauropod. One of the most fascinating aspects of sauropods is the rapid growth they achieved to reach their enormous sizes. A sauropod egg, which was about 25 cm in length, hatched into a juvenile that grew significantly within ~7 years, and it took another 7 years for the juvenile to mature into an adult. This swift growth was likely driven by various environmental factors, including the need to outgrow predators and the necessity to reach a size that would allow access to different food sources. When examining sauropods, it's essential to consider the concepts of isometric and allometric scaling. Isometric scaling, seen in animals like frogs, means that all parts of the body grow at the same rate, maintaining the same proportions throughout life. In contrast, allometric scaling involves differential growth rates, allowing certain body parts to grow faster than others to meet specific functional needs. This is crucial for larger animals like sauropods to overcome the constraints imposed by the square-cube law, where doubling in size increases surface area fourfold and volume eightfold.

Cope's Rule, named after the paleontologist Edward Drinker Cope, is a hypothesis that suggests animal lineages tend to evolve toward larger body sizes over geological time. This idea, formulated in the late 19th century, is based on observations of fossil vertebrates, particularly dinosaurs. Cope noticed a trend of increasing body sizes in successive generations within certain lineages. This concept of "phyletic gigantism" proposes that as species evolve, they generally increase in size, which can offer various advantages such as enhanced predatory abilities, improved defense mechanisms, and better resource acquisition. The principle highlights a trend where larger size within a lineage can provide a competitive edge in survival and reproduction, leading to a gradual increase in size over evolutionary time.

Larger body sizes can confer several benefits. For instance, bigger animals may have access to larger territories, more substantial prey, and better mating opportunities. They may also experience increased success in both intraspecific and interspecific competition. Additionally, larger size can lead to extended longevity, higher intelligence due to increased brain size, greater thermal inertia, and improved survival during periods of resource scarcity or climatic extremes. These advantages can create an "evolutionary arms race," where species continuously adapt to outcompete each other, driving the progression toward larger sizes over time. For example, larger predators can tackle bigger prey, while larger herbivores like sauropods can reach higher vegetation and store more fat reserves to survive harsh conditions.

The benefits of larger body size are numerous, as highlighted by Hone and Benton. These include defense against predation, increased success in predation, a broader range of acceptable food, and improved mating success. Larger animals are more successful in intraspecific competition (competition within the same species) and interspecific competition (competition between different species), tend to have longer lifespans, possess higher intelligence (correlated with larger brain sizes), have better thermal inertia (ability to maintain a stable internal temperature), and can survive through lean times and climatic variations more effectively. These factors contribute to the evolutionary success of larger animals, despite the associated costs. This may be a large reason contributing to the success of sauropods.

However, Cope's Rule is not without its controversies and exceptions. While many lineages do exhibit size increases, some remain stable or even decrease in size over time. Factors such as environmental changes, competition, and evolutionary constraints can influence the direction of size evolution within a lineage. The fossil record might also be biased toward preserving larger species due to their greater likelihood of fossilization, potentially skewing our understanding of size evolution. Additionally, larger animals face significant challenges, including the need for more resources, slower reproduction rates, and higher risks of extinction during environmental changes. The trade-offs associated with larger size mean that not all lineages follow this trend uniformly, and many species may evolve different strategies to survive and thrive.

The drawbacks of larger body size of a sauropod also need to be considered. These include an increased vulnerability to predation during development, higher demands for energy and resource (more evidence for endothermy in sauropods?), and the risk of extinction, which bring significant challenges. Larger animals typically have longer generation times, which means slower rates of evolution and reduced ability to adapt quickly to changing environments. They also have lower abundance due to smaller gene pools and lower fecundity, meaning they produce fewer offspring. These disadvantages highlight the balance of evolutionary pressures that shape the size and survival strategies of different species. Thus, although larger body size offers numerous benefits, it also comes with significant risks and challenges that must be managed through various adaptive strategies.

Even sauropods get sick

One item that until now is exceptionally rare in the fossil record is to see evidence of infections or parasites in sauropods, but no doubt had a significant role in shaping the sauropod species adaptions and radiation. The sauropods were constantly exposed to bacteria, viruses, fungi, and parasites as modern animals are currently exposed to. Living in herds could have made them especially susceptible to "catching colds" or the likes.

Recent research by Woodruff et al. has revealed the potential for infections in the bones of sauropods, shedding light on two significant aspects: the susceptibility of sauropods to infections and the pneumatization of bones linked to their respiratory system. The study focused on bone lesions in a previously identified sauropod fossil, called MOR 7029, to determine their causes. Unlike veterinary necropsies, these paleontological studies faced challenges due to the preservation of permineralized bone, which can obscure understanding of the pathology. The location of these lesions aids in differential diagnosis, with considerations such as healed fractures or arthritis. However, given the unique pneumatic bone structure of sauropods, the focus shifted to potential pulmonary issues. The lack of soft tissue preservation makes precise diagnosis difficult, but three main possibilities were considered, including neoplasia, airsacculitis, and mycobacteriosis.

Neoplasia, including air sac carcinoma or pulmonary neoplasia, was considered but did not match the lesion patterns in MOR 7029. Mycobacteriosis, common in birds, also presented inconsistent lesions. For that reason, our focus is centered on airsacculitis, considering its association with osteomyelitis and its prevalence in birds. Diseases like chlamydiosis, aspergillosis, and pneumoconiosis were also evaluated. Ultimately, airsacculitis with associated osteomyelitis was tentatively proposed as the most likely explanation due to its consistency with the observed lesions. The presence of lesions in adjacent vertebrae suggests infection through an extensive network of pneumatic diverticula, similar to features in modern birds, providing insight into respiratory infections in sauropods. MOR 7029 represents a groundbreaking case of avian-like airsacculitis in a nonavian dinosaur, marking the first instance of such a lesion in this group. This tentative diagnosis offers significant insights into ancient diseases and sauropod physiology (Fig. 2.3). We will focus on the respiratory system in the next chapter, but the importance of a respiratory infection in sauropods presents a new, but very logical way of viewing sauropods interacting with their environment.

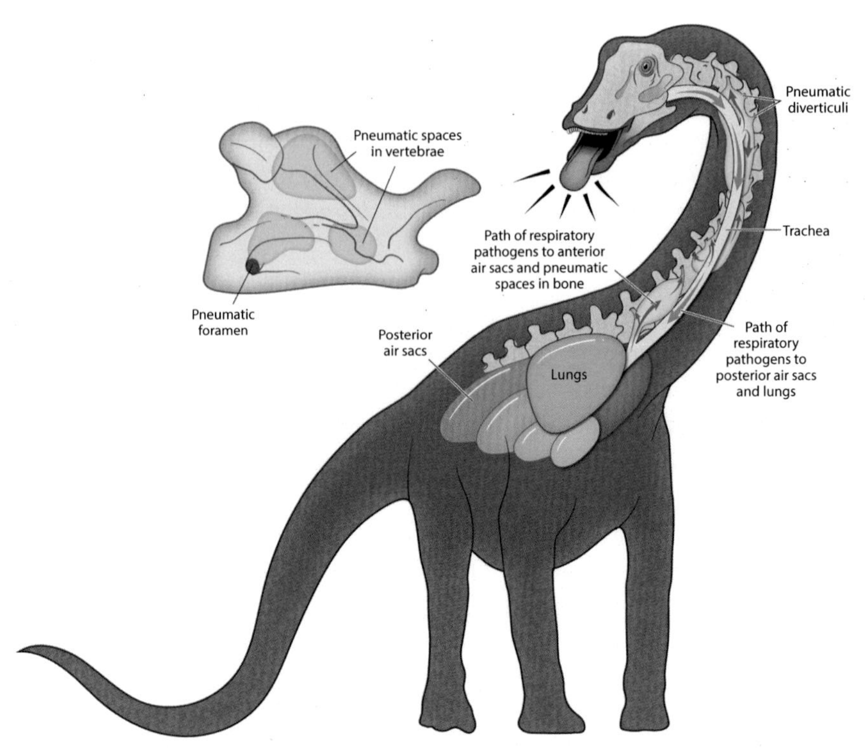

Fig. 2.3 Possible infection of a sauropod due to possible integration of respiratory system with bones. *(Based on Woodruff DC, Wolff EDS, Wedel MJ, Dennison S, Witmer LM: The first occurrence of an avian-style respiratory infection in a non-avian dinosaur, Sci Rep 12:1954, 2022.)*

Social habits

Multiple lines of fossil evidence, including bone beds and trackways, suggest that sauropods were social animals that congregated in herds. These herds varied in composition among different species. For instance, some bone beds, like the one found in the Middle Jurassic of Argentina, indicate herds consisting of individuals from different age groups, including both juveniles and adults. Conversely, several other fossil sites and trackways suggest that many sauropod species preferred to travel in herds segregated by age, with juveniles forming distinct groups separate from adults. This segregated herding behavior has been observed in species such as *Alamosaurus*, *Bellusaurus*, and certain diplodocids. The evidence of these different herding strategies points to a complex social structure within sauropod communities,

Fig. 2.4 Age-segregated sauropod herds with different skin colors based on their environment.

reflecting varying needs and behaviors across species and developmental stages (Fig. 2.4).

In a comprehensive review by Myers and Fiorillo, the authors attempted to elucidate the reasons behind the apparent prevalence of segregated herds among sauropods. Studies on microscopic tooth wear patterns revealed differences in the diets of juvenile and adult sauropods, suggesting that herding together may not have been as efficient as herding separately, where members could coordinate their foraging activities more effectively. Juveniles, with their smaller size and different dietary needs, might have benefited from different types of vegetation compared to adults. Additionally, the significant size disparity between juveniles and adults likely influenced their feeding and herding strategies. Larger adults may have required different foraging areas and resources than smaller, growing juveniles, leading to the formation of age-segregated groups.

This fits well with evidence of sauropods being "precocial," or a state of development in which a young animal is relatively mature and mobile from the moment of birth or hatching. Precocial species are characterized by offspring that are born or hatched with their eyes open, covered in down or fur, and capable of moving around and often even feeding themselves shortly after birth. This contrasts with "altricial" species, where the young are born in a much more helpless state, requiring significant parental care and feeding. In the context of precocial animals, the young are typically well-developed with fully functional senses and limbs. This early independence allows them to evade predators more effectively and reduces the immediate burden on the parents for extensive caregiving. Examples of precocial species include many ground-nesting birds like ducks and chickens, as well as some mammals like hares and guinea pigs.

Given that the segregation of juveniles and adults likely occurred shortly after hatching, coupled with evidence suggesting that sauropod hatchlings were precocial, Myers and Fiorillo concluded that species exhibiting age-segregated herds probably did not engage in extensive parental care. Precocial young are relatively mature and mobile from the moment of birth, which would have allowed juvenile sauropods to fend for themselves more quickly, reducing the need for prolonged parental care. Conversely, researchers studying age-mixed sauropod herds proposed that these species may have provided care for their young over an extended period before reaching adulthood. It has been estimated that the incubation period for sauropod eggs was likely between 65 and 82 days, suggesting a relatively short period of parental investment before the young were independent.

The extent to which herding behaviors varied between segregated and age-mixed groups across different sauropod taxa remains uncertain. Discovering further instances of gregarious behavior among different sauropod species is crucial for identifying potential distribution patterns. Understanding these behaviors can provide insights into the social structures, survival strategies, and ecological roles of sauropods. As more fossil evidence is uncovered and analyzed, it may become possible to draw clearer distinctions between the social dynamics of different sauropod species and how these behaviors contributed to their success and adaptation in various prehistoric environments.

Thick-skinned sauropods

The thickness of sauropod's skin has been a subject of considerable interest and study among paleontologists. Although direct fossil evidence of sauropod's skin is relatively rare, there are some insights based on skin impressions and comparisons with modern animals. Generally, sauropod's skin was thick and tough, providing significant protection against environmental hazards, minor injuries, and possibly even predator attacks. This robustness would have been essential for their survival, considering their massive size and the diverse environments they inhabited. The thickness of the skin likely varied across different parts of the body. For example, the skin over more vulnerable areas like the belly might have been thicker compared to the skin on less exposed areas.

Fossilized skin impressions from sauropods provide some clues about their skin's texture and thickness. These impressions often reveal a pattern of scales or tubercles, indicating a rugged and possibly multilayered structure. By comparing these skin impressions and known fossil evidence with modern animals like elephants and large reptiles, scientists infer that sauropod's skin could have been several centimeters thick. For instance, the skin of modern elephants can be up to 2.5 cm thick, suggesting that sauropod's skin might have been similar or even thicker in some areas. This comparison provides a valuable reference point for understanding the possible thickness of sauropod skin.

Sauropods' thick skin served several critical functions. The primary function was to provide a robust protective barrier against physical injuries, parasites, and environmental challenges like thorny vegetation. In addition to protection, the skin's structure, including its thickness, played a crucial role in thermoregulation. Thick skin can help retain heat, which would have been beneficial in cooler climates or especially during the night.

Osteoderm is an interesting development as a unique adaption that is generally not observed until the titanosaurs clade of sauropods, which were the "youngest" on an evolutionary scale before the Chicxulub impact. Osteoderm refers to the skin that has a bone developed within it. At the cellular level, this has been demonstrated to occur through a process known as epithelial-to-mesenchymal transition within the dermis. The bone pieces that form in the dermis are not generally connected and are interspersed within the skin. Sometimes, they protrude through the skin to form spikes, sometimes they become hardened like a club, and other examples can be more like fins. Common examples of these osteoderm developments are seen in the stegosaurs and *Ankylosaurus* dinosaurs. It isn't clear why the titanosaurs developed the osteoderm independently of other sauropod clades, but there are some interesting possibilities to consider.

One hypothesis is that osteoderms in titanosaurs provided a form of protection against predators. The heavily armored skin would have made it difficult for predators to penetrate and cause injury. This defensive adaptation could have been particularly advantageous, given the size and relatively slow movement of these massive dinosaurs, which might have made them more vulnerable to attacks. Additionally, the presence of osteoderms might have helped in thermoregulation. The bony structures could have acted as radiators, helping to dissipate heat in the hot climates that these dinosaurs often inhabited. The vascularization within the osteoderms could have facilitated this process, aiding in maintaining optimal body temperatures. This is similar to what has been hypothesized with the neck when we had discussed whether sauropods were exothermic or endothermic.

Evidence for sauropod's skin pigmentation is limited and primarily inferential, as direct evidence like fossilized skin pigments is extremely rare. Direct fossil evidence of skin pigmentation in sauropods is currently lacking. Unlike some smaller dinosaurs and prehistoric birds, where melanosomes (pigment-containing organelles) have been preserved in fossilized feathers, similar preservation has not been found in sauropod's skin fossils. Fossilized skin impressions provide information about the texture and structure of the skin but do not preserve the actual pigments. These impressions show patterns of scales and potentially other skin features, but the color and pigmentation cannot be determined directly from these fossils.

Indirect evidence and comparisons with modern animals offer some insights. Modern reptiles, such as lizards and crocodiles, display a wide range of skin colors and patterns that serve various functions, including camouflage, temperature regulation, and social signaling. Birds exhibit diverse and complex color patterns. Based on these comparisons, it is plausible that sauropods might have had some degree of skin pigmentation, possibly for similar functional purposes. However, without direct evidence, the exact nature and extent of this pigmentation remain speculative.

The environments sauropods lived in could provide some clues about their potential skin pigmentation. In open, semiarid landscapes similar to modern savannahs, camouflage would have been beneficial. This could suggest that sauropods might have had muted, earth-toned colors to blend in with their surroundings. Conversely, in more forested or varied environments, they might have displayed different pigmentation patterns to suit their specific habitats.

Hypotheses about sauropod's skin pigmentation include various functional considerations. Camouflage would have been a significant evolutionary advantage for sauropods, particularly for juveniles and smaller individuals. Earth tones and patterns that broke up their outline could help them avoid predators. Pigmentation can also play a role in thermoregulation. Darker colors absorb more heat, which could be advantageous in cooler climates or for individuals needing to warm up quickly. Lighter colors reflect heat, which could help prevent overheating in hot climates. If sauropods had the ability to change color or had distinct pigmentation patterns, these could have been used for social signaling, such as attracting mates, establishing dominance, or coordinating movements within herds.

While there is no direct fossil evidence for sauropod's skin pigmentation, comparisons with modern reptiles and birds, as well as considerations of their environments and potential functional advantages, suggest that they could have had some form of skin pigmentation (Fig. 2.4).

Moving the herd forward...

In the last few chapters, we have been focused on what we know, or can strongly surmise, based on hard evidence from fossils, animals tracks, paleobotany, and geology. From this point, we move much more of the unknown in terms of the core physiology that a sauropod may have possessed. We take with us knowledge about the environment sauropod thrived, as well as how they moved, and their bony structures. Using what we know about homeostasis and all of which that entails, we will try to layer that onto the sauropod. We will utilize a comparative animal physiology from animals millions of year removed to see how some of their adaptions may have also been utilized by sauropods, all in an attempt to determine sauropod's physiological homeostasis or balance.

References and further reading

Bakker RT: Dinosaur feeding behavior and the origin of flowering plants, *Nature* 274:661–663, 1978.

Barrett PM, Willis KJ: Did dinosaurs invent flowers? Dinosaur-angiosperm coevolution revisited, *Biol Rev Camb Philos Soc* 76:411–447, 2001.

Chiarenza AA, Mannion PD, Farnsworth A, Carrano MT, Varela S: Climatic constraints on the biogeographic history of Mesozoic dinosaurs, *Curr Biol* 32:570–585, 2022.

Cope ED: The relation of animal motion to animal evolution, *The American Naturalist* 12:40–48, 1878.

Curry Rogers KA, Wilson JA, editors: *The sauropods*, Berkley and Los Angeles, California, 2005, University of California Press.

Fu Q, Diez JB, Pole M, et al.: An unexpected noncarpellate epigynous flower from the Jurassic of China, *Elife* 7:e38827, 2018.

Hallett M, Wedel MJ: *The sauropod dinosaurs*, Baltimore, Maryland, 2016, John Hopkins University Press.

Hone DWE, Benton MJ: The evolution of large size: how does Cope's rule work? *Trends Ecol Evol* 20:4–6, 2005.

Klein N, Remes K, Gee CT, Sander PM, editors: *Biology of the sauropod dinosaurs*, Indianapolis, Indiana, 2011, Indiana University Press.

Myers TS, Fiorillo AR: Evidence for gregarious behavior and age segregation in sauropod dinosaurs, *Paleogeo, Paleoclim, Paleoeco* 274:96–104, 2009.

Pandolf KB, Sawka MN, Gonzalez RR, editors: *Human performance physiology and environmental medicine at terrestrial extremes*, Traverse City, Michigan, 2001, Cooper Publishing Group. (Reprinted).

Seymour RS: Cardiovascular physiology of dinosaurs, *Physiology (Bethesda)* 31:430–441, 2016.

Woodruff DC, Wolff EDS, Wedel MJ, Dennison S, Witmer LM: The first occurrence of an avian-style respiratory infection in a non-avian dinosaur, *Sci Rep* 12:1954, 2022.

CHAPTER THREE

The sauropod's deep breath

Many of my friends who study the cardiovascular system may send me devil eyes for writing this, but the red blood cell (RBC), also known as an erythrocyte, is probably the most important cell in our body. Essentially, the next three chapters are dedicated to how the RBC takes up oxygen (respiratory), how it is transported (the vasculature), and how it gets pushed around (the heart).

Why spend so much time on this in relation to sauropods? The answer is that the cardiovascular system, in general, links all animal physiology together. When distilled even further, the three processes of red blood cell activities are very much a lynchpin for nearly all other physiologies in an organism. The cardiovascular system governs the rate at which an animal can find food (oxygen supply for the brain), can walk to and eat the food (oxygen supply for the muscle), and can digest the food for metabolic needs (oxygen supply for gut peristalsis and mitochondrial metabolism), removal of metabolic waste (extract the CO_2 waste), bone growth (flow of blood for oxygen delivery), reproduction (blood flow), and overall growth (oxygen for metabolism).

The importance of red blood cell biology is magnified for the sauropod, because of the extreme environments of the Mesozoic. Understanding how RBCs work within their bodies is key to understanding the whole of their physiology and the wonders of these ancient giants.

We will start by examining how sauropods may have extracted oxygen—a difficult endeavor, considering the high CO_2 and the comparatively low oxygen in the atmosphere of the Mesozoic, as previously discussed. However, the sauropods must have been pretty good at extracting oxygen, considering their adaptive radiation, body size, and geographic range. So how may they have done this? Let's begin by considering what respiration is, how red blood cells work, and how respiration controls blood chemistry and, last, make an assumption on the way how sauropods extracted the limited oxygen they had access to.

The delivery of oxygen from the atmosphere to the tissues inside an animal is one of the most important components of life. This is because oxygen

Balancing a Sauropod
https://doi.org/10.1016/B978-0-12-823303-0.00004-5

(O_2) is the main element used for respiration—not the act of breathing, but respiration at the cellular level. This process occurs in cells at the mitochondria where oxygen is converted into energy for the cells to function. In almost all animals besides insects, blood vessels that deliver oxygen are never more than 10 μm, or about 1/10 of 1 cm, away from the tissue. Without oxygen, cells become hypoxic, and this spreads from cell to cell, until the tissue becomes hypoxic as well and begins to die. Each of the cells in the hypoxic tissue is unable to produce enough energy, in the form of ATP, for the cell to properly function: everything—from movement of solutes out of the cells to maintaining a proper cellular membrane to keep the cell alive. As time goes on and the lack of oxygen continues, the tissue, whichever it may be, will cease to function correctly, to the point where the animal can no longer live. Thus, moving oxygen into the tissue is perhaps one of the most important processes an animal undertakes for its survival. And with the oxygen that the cells of the body demand, they must remove the CO_2, the by-product of cellular respiration. Thus, there must be a constant exchange of O_2 and CO_2 in the body. As this is multiplied in huge multicellular organisms, from man to sauropod, it is clear that an exchange system must take place.

In this section, we will focus on (1) how oxygen gets into the sauropod from the hypoxic air and (2) how CO_2, the by-product of cellular respiration, must get out of the sauropod; further, because of the CO_2-rich atmosphere, the CO_2 is buffered to minimize damage in the sauropod.

The sauropod's red blood cell

We will first examine how gases are able to move about, especially across a membrane. Fick's law of diffusion is a fundamental principle that describes the movement of particles, such as molecules or ions, from regions of higher concentration to regions of lower concentration. It quantifies the rate at which this diffusion process occurs. According to Fick's first law, the flux of diffusing particles across a unit area per unit time is proportional to the concentration gradient across that area. Mathematically, this is expressed as:

$$V_{gas} = A \times D \times (P_1 - P_2)/T$$

In this case, V_{gas} is the volume of diffused gas (e.g., O_2 or CO_2) through the tissue barrier in a particular amount of time; D is the diffusion rate of the gas, P is the pressure, T is the time, and A is the constant unique to the gas. Using this, one can get a fairly good idea that O_2 and CO_2 diffuse at similar rates across most membranes. For example, Morelli and Scholkmann's work has suggested a significant role of passive diffusion facilitated by neutral lipids, emphasizing how lipids aid O_2 in crossing biological interfaces, such as membranes and cellular boundaries. They noted the higher solubility of O_2 in lipids compared with water, aiding in O_2 diffusion in organisms like fish in cold environments. With this in mind, we will move to how O_2 and CO_2 are transported for them to be diffused across the tissue.

Remember that red blood cells (RBCs) are highly specialized cells primarily responsible for the transportation of O_2 from the lungs to the tissues and organs throughout the body, as well as the return of CO_2 from the tissues to the lungs for exhalation. In mammals, mature red blood cells are characterized by their distinctive biconcave disk shape, which increases their surface area-to-volume ratio, facilitating efficient gas exchange. This shape also enables them to deform easily, allowing passage through the smallest capillaries. Human RBCs are devoid of a nucleus and most organelles, providing maximum space for hemoglobin, the iron-containing protein that binds O_2 and CO_2. Each human RBC contains ~270 million hemoglobin molecules, and it is hemoglobin that gives RBCs their characteristic color. In mammals, the production of RBCs, known as erythropoiesis, occurs in the bone marrow and spleen and is tightly regulated by the erythropoietin hormone, which is produced by the kidneys in response to hypoxia or low O_2 levels in the blood. RBCs also play a vital role in maintaining physiological homeostasis, not only through O_2 delivery and CO_2 removal but also by contributing to the regulation of blood pH and viscosity. The lack of O_2 in the sauropod's environment and the ability of their RBCs to complete these functions efficiently are critical to understanding their physiology.

In humans, RBCs have a lifespan of about 120 days; after which, they are removed from circulation by the spleen and liver through a process called hemolysis. The breakdown of old or damaged RBCs is crucial for recycling iron and preventing the accumulation of cellular debris in the bloodstream. However, different animals have different lifespans for their nuclei. Outside of mammals, one would think most species with nucleated RBC would provide all the organelles and new transcription of hemoglobin to allow the RBC to last for quite some time. Indeed, the lack of a nucleus in RBC prevents any kind of new proteins to be made as the DNA is gone. Interestingly, nucleated RBCs are usually viable for much shorter amount of time than anucleated RBCs.

The mean corpuscular volume (MCV) of a RBC is a measure of the average volume or size of individual red blood cells (erythrocytes) in a blood sample and is a direct readout of how much hemoglobin each RBC can carry. The MCV is calculated by dividing the total volume of packed red blood cells (hematocrit) by the number of red blood cells in each volume of blood, and it is typically expressed in femtoliters. In humans, an MCV value within the normal range (80–100 fL) indicates normocytic red blood cells, whereas values outside this range indicate microcytic (smaller than normal) or macrocytic (larger than normal) cells. Microcytic cells are often associated with iron deficiency anemia or thalassemia, while macrocytic cells are indicative of vitamin B12 or folate deficiency, among other conditions.

From an evolutionary perspective, the MCV of red blood cells reflects adaptations to diverse environmental pressures and physiological demands experienced by different species. The size of RBC, as indicated by MCV, influences an organism's ability to transport oxygen efficiently, which is crucial for survival and reproduction. In mammals, including humans, MCV variations can be linked to evolutionary adaptations to altitude, climate, diet, and overall metabolic demands. For instance, species living at high altitudes, where oxygen levels are lower, might evolve larger red blood cells (higher MCV) to maximize oxygen uptake and transport. Larger RBCs have a greater surface area for oxygen exchange and can carry more hemoglobin, enhancing oxygen delivery to tissues in hypoxic conditions. Conversely, smaller RBCs (lower MCV) are advantageous in environments where rapid oxygen exchange is necessary, such as in high-metabolism species or those exposed to frequent physical exertion. Smaller cells can travel more easily through capillaries and are more efficient in exchanging gases quickly due to their higher surface area-to-volume ratio.

Dietary factors also play a role in the evolution of MCV. Species with diets rich in iron and vitamins might develop larger red blood cells, while those with limited access to these nutrients might have smaller cells, reflecting a balance between available resources and physiological needs. Overall, the evolution of MCV is a complex interplay of genetic, environmental, and ecological factors, highlighting the diverse strategies organisms have developed to optimize oxygen transport and sustain their metabolic functions in varying habitats.

Oxygen delivery to animals is equal to the cardiac output (described in the following chapter) multiplied by the oxygen content of blood (as determined by the amount of hemoglobin). In mammals, there is a linear relationship between MCV and the number of overall RBCs. For example, the smaller the RBC, the smaller the MCV, but there is an increased overall number of RBC; conversely, the larger the size of the RBC, the higher the MCV (of course!), but the fewer overall number of red blood cells. In the end, this achieves the same result in an attempt to achieve the perfect ratio of hemoglobin to hematocrit to achieve the best oxygen delivery (~15 g/DL hemoglobin and ~45% hematocrit). The importance of this ratio is it balances the benefits of hemoglobin on oxygen-carrying capacity, with the viscosity of the blood (slower blood, less likely to get it oxygenated or remove waste products).

Let's look at the most extreme examples of RBC in species living today and how that might apply to sauropods (Fig. 3.1). The smallest RBCs are from mouse deer (*Tragulus kanchil*), which are small, primitive ungulates

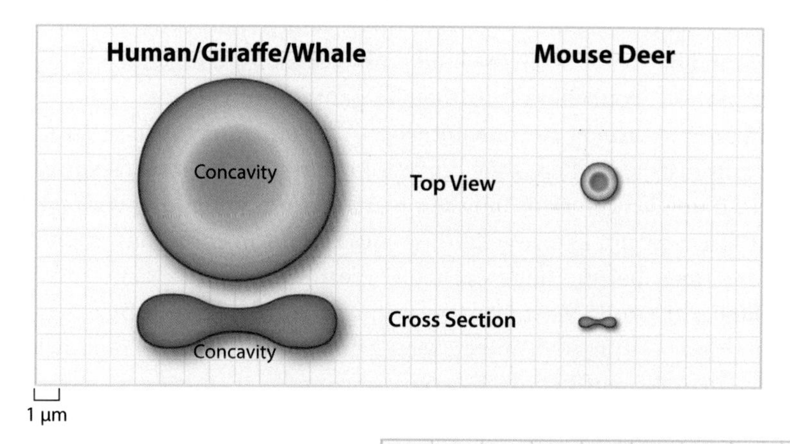

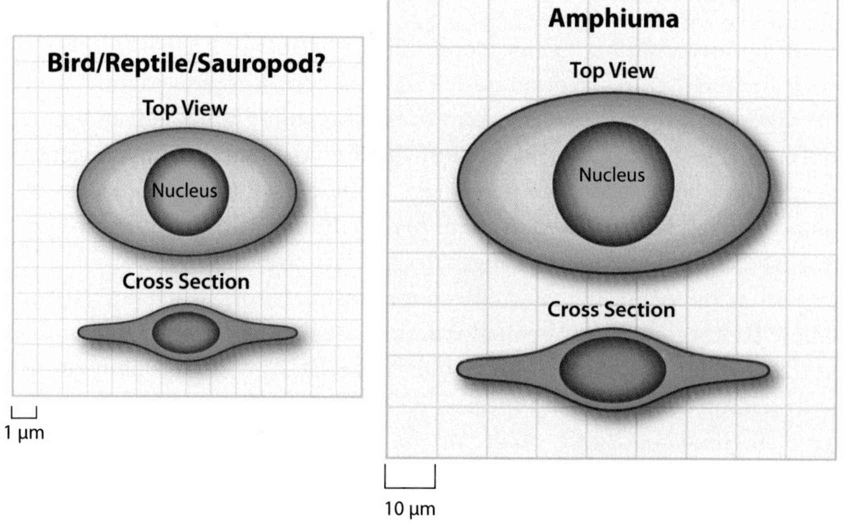

Fig. 3.1 Size of red blood cells across species, with and without a nucleus.

found in the dense forests of Southeast Asia, South Asia, and parts of Africa. Despite its name, it is not a true deer but shares some superficial similarities with both deer and mice, such as its small size and delicate features. The mouse deer typically measures about 45 cm in length and weighs around 2 kg. Their RBCs are a maximum of 2 μm in diameter and like most mammals, like a nuclei. The animals have exceptionally high metabolic rate that may indicate their need to acquire O_2 very quickly. The largest RBCs are from the amphiuma, a genus of aquatic salamanders native to the Southeastern United States, known for their eel-like appearance and adaptations to a life spent mostly in water. These slender, elongated amphibians can grow up to 1 m in length, making them one of the largest salamanders in North America. They possess tiny, vestigial limbs with reduced digits—two on the front and one to three on the hind limbs—which are functionally useless for locomotion, contributing to their serpentine movement. Amphiumas are primarily nocturnal and inhabit slow-moving waters such as swamps, ponds, and marshes, where they burrow into mud and debris. Their environment is not oxygen rich, and their metabolism is quite low; with nucleated RBC at ~60 μm in diameter, they could retain an enormous amount of O_2 at one time, but the huge size likely creates a slower velocity and certainly higher friction. Most birds, lizards, and reptiles have RBCs that are nucleated and ~10 μm in diameter. Interestingly, they all have a slightly longer shape than the perfect circles seen in humans and mammals. This doesn't appear to take into account their metabolic profile, as reptiles and birds are vastly different in metabolic rate.

A last measurement to consider in RBC physiology is the mean corpuscular hemoglobin concentration (MCHC), a hematological parameter that measures the average concentration of hemoglobin within a given volume of RBC. The MCHC is calculated by dividing the hemoglobin concentration by the hematocrit (the percentage of blood volume occupied by red blood cells) and is expressed in grams per deciliter (g/dL). In humans, a normal MCHC value typically ranges from 32 to 36 g/dL. A lower MCHC, known as hypochromia, suggests conditions like iron deficiency anemia, where RBCs contain less hemoglobin than usual. Conversely, a higher than normal MCHC, or hyperchromia, indicates RBCs are more densely packed with hemoglobin. Camelid species, including camels, llamas, and alpacas, have an elongated RBC, similar in size to that seen in birds and reptiles, and exceptionally high MCHC at around 40–50 g/dL. This makes sense in terms of the low O_2 environment these animals evolved in. The consistency of the size, shape, and presence of a nucleus across the evolutionary scale in birds and mammals likely places the sauropod RBC in the category of the size of RBCs birds and lizards also maintained. In addition, the consistency of an elongated shape of RBC to achieve higher MCHC in birds and lizards, as well as in mammals that evolved in low O_2 environments, makes it even more likely that the sauropods had a similar size of RBC to modern-day birds and lizards. Once more, it's likely that any increase in the RBC, to achieve higher O_2-carrying capacity, would likely alter viscosity of blood and blood pressure, something the sauropod was likely already fighting (see next chapters).

Acids and bases

It might be surprising to learn that the respiratory system plays a crucial role in maintaining acid-base balance in animals. This is especially important in hypoxic conditions, like the atmosphere in which sauropods lived and thrived in. Before we get into that, first, we should define what acids and bases are and how they work. Acids and bases can be defined in various ways, but physiologically, an acid is best described as a substance that donates hydrogen ions, while a base is one that accepts hydrogen ions. A buffer, typically a combination of a weak acid and its conjugate base in aqueous solution, helps stabilize hydrogen ion concentrations when strong acids or bases are introduced. The pH scale measures hydrogen ion concentration, based on the concentration in pure water, which is 1.0×10^{-7} mol/L. Solutions with hydrogen ion concentrations greater than 10^{-7} mol/L are acidic, and those with lower concentrations are basic (alkaline). For instance, the gastric juice in the stomach has a hydrogen ion concentration of about 10^{-1} mol/L (pH 1), while pancreatic secretions are around 10^{-8} mol/L (pH 8). The pH of arterial blood is maintained within a narrow range of 7.35–7.45, with an optimal value of 7.4, which is critical because hydrogen ions are highly reactive and can interact with negatively charged regions of molecules. This interaction can significantly alter protein structures and functions, such as how hemoglobin's ability to bind oxygen is affected by pH changes (Bohr effect, as described earlier).

Because of all the reactivity with hydrogen ions, there must be buffers to regulate this, and in general, there are three main buffers in animals, including bicarbonate, phosphate, and proteins. Each of these reacts in some way with the CO_2 produced as a by-product of cellular respiration. As a by-product waste that needs to be removed, it is not surprising therefor that CO_2 is up to 20 times more soluble in blood as O_2.

Starting with bicarbonate buffering, it acts as a buffer pair of the weak acid, carbonic acid, and its conjugate base, bicarbonate, both of which are formed by CO_2. In mammals, ~90% of CO_2 is carried as bicarbonate ions due to the following reaction:

$$CO_2 + H_2O \longleftrightarrow H_2CO_3 \longleftrightarrow H^+ + HCO_3^-$$

The CO_2 combines directly with water molecules (H_2O) and forms carbonic acid ($H_2CO_3^-$), as described in the Bohr and Haldane reactions. The carbonic acid is unstable and quickly disassociates into hydrogen ions (H^+) and a bicarbonate ion (HCO_3^-). There is not a lot of carbonic acid that is formed just with water and CO_2, but that is catalyzed quickly to carbonic acid by the presence of the ancient enzyme carbonic anhydrase. However, carbonic anhydrase is not found in plasma and is localized predominantly in red blood cells, which makes the reaction between CO_2 and H_2O go up to 13,000 times faster. The presence of carbonic anhydrase in RBC is therefore an important buffering mechanism in the blood, and in a high CO_2 atmosphere, and an enormous circulation with long delays between production of CO_2 and removal, it is possible sauropods needed to make this enzyme in their RBC in large quantities.

The ability of the bicarbonate system to function as a buffer of fixed acids in animals is due to the ability of the respiratory system to remove carbon dioxide from the body. At a temperature of 37°C, there is ~1000 times as much CO_2 present dissolved in blood plasma as there is carbonic acid; however, the dissolved carbon dioxide and carbonic acid are in chemical equilibrium with each other. Therefore, what would happen with increased breath cycles (i.e., hyperventilation) or decreased breath cycles (i.e., hypoventilation)? Hyperventilation of a sauropod would cause blood to become more basic (alkalosis) because the CO_2 and carbonic acid would be decreased in the blood due to excessive exhalation of the CO_2. Alkalosis occurs in animals at high altitudes due to the lack of oxygen and the need to continue to try to take in air to extract the oxygen. It is highly unlikely the sauropod regularly hyperventilated during homeostasis due to the success of the sauropod, and so other mechanisms likely ensured this wasn't an issue (e.g., the oxygen disassociation curve described earlier or other mechanisms described in the following). The chemical reaction would shift the equilibrium to favor the more basic bicarbonate. Conversely, hypoventilation would cause the blood to become more acidic (acidosis) because the CO_2 and carbonic acid would be increased in the blood and shift the equilibrium, according to the previous equation, in the opposite direction.

As the pH of blood is so tightly regulated across species in the small 7.4 window, regulating these processes in a sauropod in a high CO_2 or low O_2 environment would be paramount.

Because maintaining blood pH is so vital, buffers are not the only way in which this is controlled. Like most physiology, there are negative feedback loops built into animals to ensure blood pH is brought back to normal. This is generally done by unique receptors on specialized cells that are present to "sense" the environment. The receptors are called chemoreceptors, which are sensory receptors that detect chemical stimuli in the environment or within the body. These receptors are crucial for processes such as taste, smell, and the regulation of blood chemistry. They are found in various locations, including the taste buds of the tongue and the olfactory receptors in the nasal cavity. Important for this discussion, these types of receptors are also present on the carotid and aortic bodies in the cardiovascular system. Chemoreceptors work by binding specific molecules such as those buffering blood pH and triggering a cascade of cellular events that result in a nerve impulse sent to the brain, where the chemical information is processed and interpreted. This allows organisms to respond to changes in their internal and external environments, ensuring proper physiological function and adaptation. In the case of respiratory systems, a case of blood acidosis would cause the sauropod to increase breathing rates and blood alkalosis to slow down breathing.

As noted above, the reaction to produce the bicarbonate and carbonic acid primarily occurs in RBC where carbonic anhydrase is predominantly localized. In the view of location is everything, the addition of hemoglobin to this reaction creates an even stronger buffer—with RBC being the primary site of hemoglobin creation, we have red blood cells creating an extra-strong buffering system. This allows bicarbonate concentrations to rise when there is more hemoglobin present as more carbonic acid is created (by the addition of more CO_2). The increase in bicarbonate concentrations is greater when there are more hemoglobin because, as CO_2 is added to the blood, the hydrogen ions formed by the dissociation of carbonic acid are buffered by hemoglobin.

Proteins, in general, also play a role in the blood acid-base buffering, specifically a part of proteins called imidazole groups. In chemistry, imidazole groups refer to the functional groups containing the imidazole ring, a five-membered aromatic heterocycle with two nitrogen atoms at nonadjacent positions. The imidazole ring is a common structure in many biologically active molecules, especially the amino acid histidine. It features a planar structure with delocalized electrons, contributing to the imidazole aromaticity. The nitrogen atoms in the imidazole ring can act as both proton donors and acceptors, making this group versatile in catalyzing biochemical reactions and forming complexes with metal ions. Imidazole groups are known for their significant role in enzyme active sites, where they participate in acid-base catalysis and stabilization of transition states during enzymatic reactions. Of particular importance is hemoglobin, again enriched in RBC. Hemoglobin proteins have a significant number of histidine amino acids that are altered depending on whether hemoglobin is oxygenated or deoxygenated. For example, as red blood cells loaded with oxygenated hemoglobin enter the capillaries, and oxygen disassociates from hemoglobin to become deoxyhemoglobin, imidazole groups remove hydrogen ions from the red blood cells, allowing more CO_2 to be taken up and transported as bicarbonate.

To conclude the discussion on plasma buffering, the last component of plasma buffering is phosphates. There are many organic phosphates that can act as buffers, including glucose-1-phosphate and adenosine trisphosphate, which act via the chemical reaction, with PO_4 being the phosphate:

$$H_2PO_4 \longleftrightarrow H^+ + HPO_4{}^{2-}$$

These phosphates occur mainly as the metabolic by-products of the cellular respiration.

Oxygen absorption in an oxygen-poor environment

The next question that arises is whether a low O_2 environment, such as in the late Jurassic, would necessarily alter the way in which sauropods breathed and extracted O_2 from the environment. In essence, do we have to completely change the way we view the pulmonary system of sauropods based on the hypoxic environment they lived in? The short answer is likely no, based on what we know about oxygen saturation curves.

Oxygen dissociation curves are graphical representations that depict the relationship between the partial pressure of oxygen (pO_2) and the percentage saturation of hemoglobin with oxygen. These curves are helpful in understanding how oxygen is loaded onto and unloaded from hemoglobin, facilitating the efficient transport of oxygen from the lungs to tissues and vice versa. The typical oxygen dissociation curve has a sigmoidal (S-shaped) look on a graph, reflecting the cooperative binding nature of O_2 and hemoglobin. At low pO_2 levels, such as those found in tissues, hemoglobin binds oxygen less readily. However, as pO_2 increases, such as in the lungs, hemoglobin's affinity for oxygen increases significantly, leading to more rapid oxygen binding. This phenomenon is due to structural changes in the hemoglobin molecule as it binds oxygen, which makes it easier for subsequent oxygen molecules to attach. The steep middle portion of the curve represents this rapid loading of oxygen, while the plateau at higher pO_2 levels indicates hemoglobin nearing full saturation and binding additional oxygen molecules less readily.

Several physiological factors can influence the position and shape of the oxygen dissociation curve, shifting it to the right or left and thereby altering hemoglobin's affinity for oxygen (Fig. 3.2). A rightward shift of the curve, known as the Bohr effect (discussed more later), occurs under conditions such as increased CO_2 concentration, lower pH (acidosis), elevated temperature, and higher levels of 2,3-bisphosphoglycerate (2,3-BPG). These conditions are typically found in metabolically active tissues, where oxygen demand is high. The rightward shift facilitates the release of O_2 from hemoglobin, ensuring that active tissues receive an adequate supply. Conversely, a leftward shift of the curve indicates an increased affinity of hemoglobin for oxygen and occurs under conditions oflower carbon dioxide concentration,

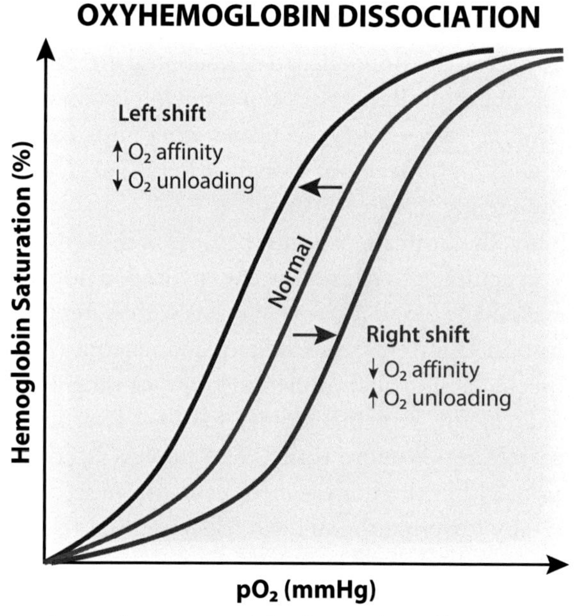

Fig. 3.2 Oxygen dissociation curve and the factors that influence it.

higher pH (alkalosis), decreased temperature, and lower levels of 2,3-BPG. This leftward shift is advantageous in the lungs, where it ensures maximal oxygen uptake despite relatively lower pO_2 levels compared to tissues.

The physiological relevance of the oxygen dissociation curve extends beyond simple oxygen transport. In the lungs, where pO_2 is high (~100 mmHg), hemoglobin binds to oxygen efficiently, becoming nearly fully saturated. This saturation ensures that oxygen is adequately loaded onto hemoglobin for transport to peripheral tissues. As blood circulates to tissues with lower pO_2 (~40 mmHg), hemoglobin releases oxygen, meeting the metabolic demands of these tissues. This mechanism is finely tuned to respond to varying oxygen demands under different physiological conditions. For example, during vigorous exercise, muscle tissues produce more CO_2 and hydrogen ions (lowering pH), which shifts the curve to the right and enhances O_2 release. Additionally, adaptations in the oxygen dissociation curve are crucial for animals living at high altitudes. At higher altitudes, lower atmospheric O_2 levels (such as that in the Mesozoic) necessitate physiological adjustments, such as increased hemoglobin affinity for O_2 and elevated RBC production, to ensure adequate O_2 uptake and transport.

As noted before, the Bohr effect is a physiological phenomenon first described by Christian Bohr in 1904, detailing how the pH of blood and the concentration of CO_2 influence the oxygen-binding affinity of hemoglobin. This effect is vital for efficient oxygen transport and delivery to body tissues. At a basic level, the Bohr effect describes how increases in CO_2 levels and decreases in pH result in hemoglobin reducing its affinity for O_2. This process facilitates the release of O_2 in tissues where it is most needed, particularly those with high metabolic activity where CO_2 levels are elevated and pH is consequently lower.

Biochemically, the Bohr effect occurs because of the sensitive response of hemoglobin's structure to changes in the hydrogen ion concentration. When CO_2 levels in the blood increase, it reacts with water to form carbonic acid, which dissociates into bicarbonate ions and protons (H^+). These protons bind to specific amino acids in hemoglobin, causing a conformational change that reduces the protein's affinity for O_2. This alteration enables hemoglobin to release O_2 more easily. The process is crucial in areas of active respiration within the body, where cells consume oxygen and produce CO_2, thereby lowering the pH and triggering the release of O_2 from hemoglobin.

In the lungs, the opposite scenario occurs, reversing the Bohr effect. Here, CO_2 is expelled, which increases the pH of the blood. The lower concentration of hydrogen ions leads to a conformational change in hemoglobin that increases its affinity for O_2, facilitating the uptake of O_2 into the blood. This enhanced oxygen-binding capacity is essential for the efficient loading of oxygen in the lungs, ensuring that sufficient oxygen is available to be transported to all parts of the body. The Bohr effect thus plays a critical role in regulating the delivery and release of oxygen in accordance with the varying metabolic needs of different tissues, optimizing respiratory efficiency and supporting cellular metabolism.

Apart from the Bohr effect is the Haldane effect which describes how O_2 levels in the blood influence the capacity of hemoglobin to bind and release CO_2. This effect is crucial for efficient gas exchange during respiration, particularly in the transportation of CO_2 from tissues back to the lungs. Essentially, when hemoglobin releases oxygen to tissues, its affinity for CO_2 increases. This relationship is particularly beneficial as it allows blood to carry more CO_2 away from the actively respiring tissues, where oxygen is being consumed and CO_2 is being produced.

At the biochemical level, the Haldane effect is explained by changes in the structure of hemoglobin as it releases O_2. Oxygenated hemoglobin has a lower affinity for CO_2 compared with deoxygenated hemoglobin. When O_2 is delivered to tissues and released from hemoglobin, the protein undergoes a conformational change that increases its ability to bind CO_2. This binding is not only for CO_2 molecules directly but also enhances the conversion of carbon dioxide into bicarbonate ions (HCO_3^-) and protons (H^+) in red blood cells. This reaction is catalyzed by the enzyme carbonic anhydrase (more on that below). Bicarbonate then diffuses into the plasma, further facilitating CO_2 transport. In the lungs, similar to the Bohr effect, the reverse process occurs with Haldane. As hemoglobin binds O_2 in the lungs, its affinity for CO_2 decreases, causing CO_2 to be released from hemoglobin. This release is facilitated by the conversion of bicarbonate back into CO_2. The decrease in CO_2-binding capacity upon oxygenation of hemoglobin helps in releasing more CO_2 into the alveoli of the lungs, from where it can be exhaled.

This cyclical interaction between O_2 and CO_2 using the Bohr and Haldane effect is essential for maintaining the balance of these gases in the bloodstream and supports the overall process of respiration and metabolic regulation. It is a process that is conserved across all living organisms that utilize hemoglobin. The hemoglobin protein has been present nearly unaltered for over 600 million years, evolving originally in single-celled organism, and thus, its presence in sauropod RBC doesn't seem out of the question and in fact was highly likely. Thus, regardless of the low O_2 environment, there was still some O_2 present in the atmosphere (as we have already noted), which was enough to keep the curve shifted to the right. Overall, however, it is unlikely sauropods had too much trouble extracting O_2 to the environment to perform physiological tasks (e.g., production of ATP via mitochondria).

How did the sauropods get that oxygen off the curve?

We next move to the process of how the O_2 and CO_2 were exchanged, in the sauropod lungs...but what kind of lungs were they? Assuming the sauropod was completely land-dwelling, they had to have had lungs to extract oxygen from the atmosphere, as opposed to gills that extract oxygen from water. In general, there are two types of lungs to consider. The first lung is mammalian with a singular inhalation and exhalation. This process works efficiently, with the air inhaled to the distal lungs where the capillaries are, the O_2 extracted and CO_2 released, and the same air is then exhaled. The potential issue with this is not all the O_2 can be absorbed, and not all the CO_2 is released; there remains dead space in the sacs of the lungs where the air is drawn into. The second type of lung is used by birds, and this is a much more efficient process for O_2 extraction. There is evidence from the Therapoda lineage, later-stage dinosaurs (~65 million years old, right before the Chicxulub meteor impact), that they had bird-like lungs, with indentations through ribs, similar to modern-day chickens. But as we described in Chapter 1, sauropods aren't necessarily along the same evolutionary path as Therapoda. However, the need of sauropods to efficiently extract O_2 from the O_2-poor environment, and the huge requirement for O_2 of a sauropod body would mean the lungs would have to be the absolute most efficient possible. For this reason, we will make the strong assumption that sauropods very likely had a respiratory system similar to birds. Their long neck may have even provided the extra space to store the anatomical components of the lungs, enabling their larger body size. So how might have these lungs worked?

Broadly, sauropods likely used air sacs that functioned like an advanced respiratory system that essentially pushed a continuous flow of air through the lungs. The air sacs acted as reservoirs, ensuring a constant supply of oxygen-rich air to the lungs and facilitating the exchange of gases. Air sacs were essentially an extension of lungs—when sauropods inhaled, fresh air would enter the air sacs, and when they exhaled, the air would pass through the lungs, allowing for efficient gas exchange. It was like having an extra set of lungs. This system also likely had a series of interconnected air sacs that extend into parts of their bones, fundamentally different from the mammalian lungs that expand and contract during breathing. Instead, these air sacs act as bellows, pushing and pulling air through the lungs in a continuous flow, ensuring a constant supply of fresh oxygen. Unlike mammals that have a two-cycle inhalation and exhalation, birds, and likely sauropods, have two inhalation and two exhalations to complete one breathing cycle (e.g., see Fig. 3.3).

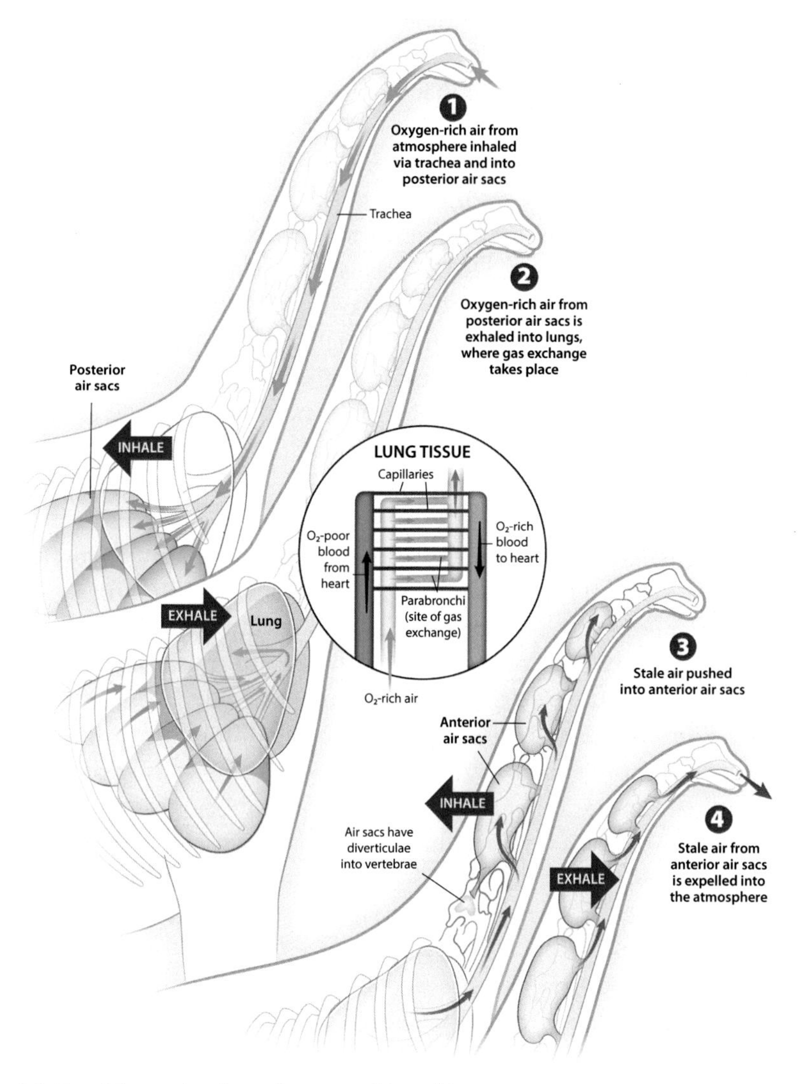

Fig. 3.3 Possible mechanism of sauropod breathing.

A distinctive feature of the avian (i.e., sauropod) respiratory system is its flow-through ventilation mechanism. Air inhaled through the trachea only moves in one direction—entering through the posterior air sacs, passing through the lungs where gas exchange occurs, and exiting through the anterior air sacs. This unidirectional airflow is more efficient than the bidirectional flow observed in mammals, as it allows for a higher rate of gas exchange and prevents the mixing of fresh and spent air.

The air sacs of a bird do not participate directly in gas exchange but are vital for ventilating the lungs. Their expansion and contraction facilitate the continuous movement of air through the respiratory system, enhancing the bird's ability to extract oxygen. The sauropod likely had many, voluminous air sacs. It is important to note that air sacs in birds are not vascularized, as you might imagine they would be in mammals, and instead are fairly acellular in nature, more likely to ensure oxygen-rich air doesn't escape in the posterior air sacs and CO_2-rich air doesn't escape during exhalation.

The mechanical aspect of avian respiration involves complex movements of the rib cage and sternum. During inhalation, the sternum moves forward and downward, enlarging the thoracic cavity and drawing air into the posterior air sacs. This mechanical process is tightly coordinated with the bird's metabolic needs, particularly during strenuous activities like flight. In the case of sauropods, this likely included movement as in walking or reaching for leaves.

One of the biggest issues of inhaling air faced by the sauropods is the distance the air needs to travel from inhalation at the nostrils to the posterior air sac (assuming avian-like lungs) in the chest, surrounding the rib cage (again, assuming a similarity with the avian like lungs). This is quite a distance for the air to be sucked into the sauropod, but the converse of this is that the volume of air would fill the gigantic posterior air sacs. The syrinx, located at the base of a bird's trachea, not only facilitates complex vocalizations but also regulates the flow of air through the airways. By adjusting the tension on the syrinx, birds can control air flow, which is crucial during singing or other vocal activities without compromising the efficiency of respiration. It is possible sauropods could also use a syrinx for communication, especially with such a long trachea to produce low-frequency sounds. It's interesting to imagine what a bass band of sauropods would have sounded like! However, to ensure there was enough air getting into the sauropod, perhaps these syrinxes were minimal, with consequences for a more contracted trachea.

When air enters the trachea, it encounters the primary site of airway resistance. The behavior of inhaled airflow resembles that of a fluid and can occur as either laminar or turbulent flow. Laminar flow is characterized by concentric layers of air moving at different velocities, much like telescoping cylinders. The layer of air closest to the tracheal wall moves the slowest due to frictional forces, while the central layer moves the fastest. In a more rigid and smooth tube, airflow behavior adheres to Poiseuille's law, which states that the pressure difference is directly proportional to the flow rate multiplied by the resistance. According to this law, resistance is directly proportional to the air's viscosity (affected by factors such as humidity and particulate content) and the tube's length. This implies that the longer the trachea, the greater the resistance, making it harder to draw air into the lungs. Additionally, resistance is inversely proportional to the fourth power of the tube's radius, meaning that if the radius is halved, the resistance increases 16-fold. This relationship underscores the significant impact of sauropod trachea length and diameter on airflow resistance (Fig. 3.4).

Turbulent flow is a type of fluid movement characterized by chaotic and irregular motion, where fluid particles follow erratic paths rather than smooth, orderly layers seen in laminar flow. This turbulent behavior typically occurs at higher velocities or when the fluid encounters obstacles or changes in direction, leading to vortices, eddies, and rapid fluctuations in pressure and velocity. The onset of turbulence is often described by Reynolds number, a dimensionless quantity that predicts the transition from laminar to turbulent flow based on fluid properties and flow conditions. In the context of airflow within the respiratory system, turbulent flow can increase airway resistance and energy expenditure required for breathing. Unlike laminar flow, where resistance is primarily influenced by viscosity and tube length, turbulent flow is heavily impacted by factors such as flow velocity and the roughness of airway surfaces. For these reason, almost regardless of the type of respiratory system the sauropod had, assuming a very long trachea from head to the base of the body, we have to imagine the trachea was a large-diameter tube to ensure as little resistance as possible in terms of laminar flow and that it was likely as smooth and unconstructed as possible to minimize any turbulent airflow down the trachea.

The uniqueness of avian respiration doesn't stop with the air sacs and four-cycle breathing. Birds and sauropods also contained pneumatized bones, as described in Chapter 1. Here, we note their relationship to breathing. This has been worked out by Dr. Markus Lambertz from the Institute

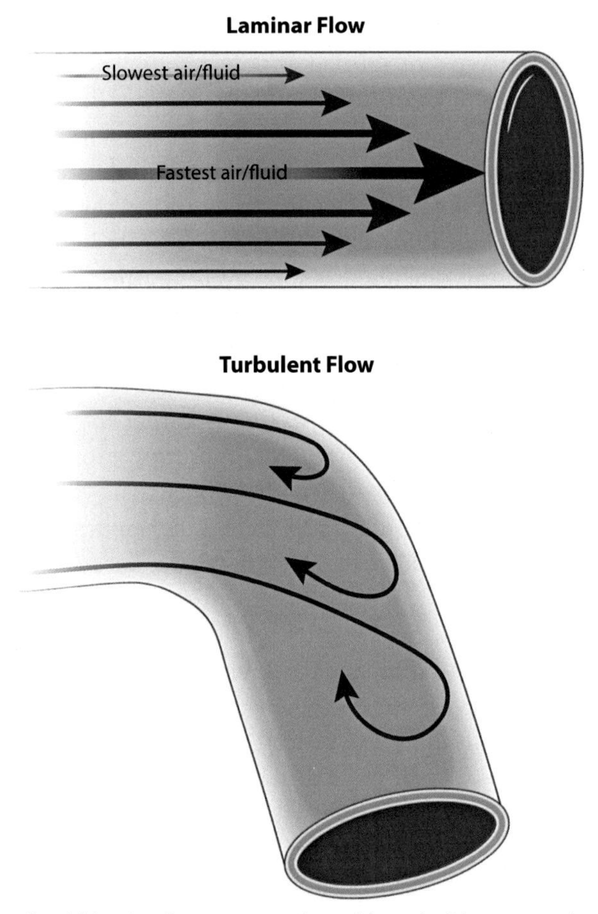

Fig. 3.4 Size of red blood cells across species, with and without a nucleus.

for Zoology at the University of Bonn in Germany. Dr. Lambertz's work explains that the avian air sacs not only acted as protrusions from the lung as described before, but they can invade the bones. One can imagine in sauropods where this would be highly advantageous to lighten the load of the neck in terms of having less heavy/dense bones to carry. Evidence from sauropods was also demonstrated by Woodruff et al., as described in Chapter 2, where a respiratory infection caused bone lesions, likely due to the connected nature of air sacs to bones of at least the vertebrate. The air sacs in the bones of birds were found to have a very fine and densely packed fibers, acting in a similar way to air sacs being acellular in nature. These structures were termed pneumosteum, and the identification could help in distinguishing similar structures in sauropod bones.

Overall, birds' adaptations to various environments, such as high altitudes, demonstrate the robustness of the avian respiratory system. Species like the bar-headed goose have evolved to have higher-affinity hemoglobin and an increased lung surface area, which helps them thrive in low-oxygen environments by maximizing oxygen uptake. Moreover, the avian respiratory system also plays a crucial role in thermoregulation, something the sauropod would be keen to keep under control as well. In birds, the air sacs help dissipate excess body heat generated during flight. This cooling mechanism is vital for maintaining an optimal body temperature and is particularly important during high-energy activities. For these reasons and as noted earlier, the likelihood is the sauropods maintained a bird-like respiratory system. We may not describe as perfectly as it was present in the sauropod, but without any kind of fossil evidence, the basics of their system are likely accurate.

Putting the sauropod respiratory system together

Between what we learned about RBC biology and the accompanying physiology that it entails, including the amount oxygen it could conceivably carry, as well as their size, a preserved RBC could provide some important indirect knowledge of sauropod physiology. There are some tangential findings on dinosaur RBC in the literature, but they are secondary to being able to accurately determine their characteristics due to dehydration and fossilization. One of our biggest conclusions is the likely presence of an avian-like respiratory system in sauropods and all of the advantages that would have entailed for moving the large sauropod body and extracting the most amount of oxygen possible from the atmosphere. With this in mind, we will move to mechanisms that transported the oxygenated RBC around the giant sauropod body.

References and further reading

Curry Rogers KA, Wilson JA, editors: *The sauropods*, Berkley and Los Angeles, California, 2005, University of California Press.

Hallett M, Wedel MJ: *The sauropod dinosaurs*, Baltimore, Maryland, 2016, John Hopkins University Press.

Hill RW, Wyse GA: *Animal physiology*, ed 2, New York, New York, 1989, Harper Collins Publishers.

Klein N, Remes K, Gee CT, Sander PM, editors: *Biology of the sauropod dinosaurs*, Indianapolis, Indiana, 2011, Indiana University Press.

Koeppen BM, Stanton BA: *Berne and levy physiology*, ed 6, Philadelphia, Pennsylvania, 2008, Mosby Elsevier.

Lambertz M, Bertozzo F, Sanders PM: Bone histological correlates for air sacs and their implications for understanding the origin of the dinosaurian respiratory system, *Biol Lett* 14:201705514 2018.

Levitzky MG: *Pulmonary physiology*, ed 6, New York, New York, 2003, McGraw-Hill.

Mohrman DE, Heller LJ: *Cardiovascular physiology*, ed 5, New York, New York, 2003, McGraw-Hill.

Morelli AM, Scholkmann F: The significance of lipids for the absorption and release of oxygen in biological organisms, *Adv Exp Med Biol* 1438:93–99, 2023.

Pandolf KB, Sawka MN, Gonzalez RR, editors: *Human performance physiology and environmental medicine at terrestrial extremes*, Traverse City, Michigan, 2001, Cooper Publishing Group. Reprinted.

Woodruff DC, Wolff EDS, Wedel MJ, Dennison S, Witmer LM: The first occurrence of an avian-style respiratory infection in a non-avian dinosaur, *Sci Rep* 12:1954, 2022.

CHAPTER FOUR

Moving blood through a sauropod: The vasculature

We move from obtaining oxygen and placing it on red blood cells to moving it around the tissue for oxygenation. In large multicellular organisms, the vasculature is a closed circuit, and the sauropod was, very likely, no different. We begin by understanding the basics of the vasculature which has a strong similarity with reptiles, birds, and mammals. We will take this knowledge and begin to layer on how the sauropod's circulatory system may have been modified, or more similar to what we know today.

In this chapter, we need to dive deep into the vascular system of sauropods because of the mystery of their long neck. The vast size of their neck and the distance between their brain and heart beg the question of how they regulated blood pressure and the overall vasculature—this may resemble challenges that a giraffe currently faces, but magnified 10 times by their enormous size and even longer neck. What we do know is the law of physics is the same today as they were 100 million years ago. The physics that regulate your blood pressure regulated the blood pressure of the sauropod—let that sink in. We need to understand the laws of vascular movement, so we can understand how the sauropod survived and thrived. Applying these rules to what we know about their extreme physiology truly elucidates what magnificent creatures they were. The question begging itself to be answered is how they survived blood pressure rates that would send our physicians into delirium. For that reason, we now move into vasculature physiology 101.

The power of two...or three

The two main cell types in the vasculature are endothelium and smooth muscle cells, and they are layered on top of each other, both acting independently and as a syncytium, to regulate vascular function. Endothelial cells form the inner lining of blood vessels, including arteries, veins, and capillaries, as well as the lymphatic vessels, and are collectively known as the endothelium. Structurally, endothelial cells are thin, flat, and polygonal in shape by being stretched by the laminar flow of blood. The cells are closely

Balancing a Sauropod
https://doi.org/10.1016/B978-0-12-823303-0.00006-9

connected to each other through proteins called tight junctions that, as the name implies, keep the cells very close together. These tight junctions between endothelium control the permeability of the vessel wall, thus regulating the exchange of substances between the bloodstream and surrounding tissues. This is especially apparent in capillaries.

Functionally, endothelial cells are integral to the regulation of blood flow and vascular resistance. They produce and release vasoactive substances such as nitric oxide, prostacyclin, and endothelin, which modulate whether the blood vessel opens (vasodilation) or constricts (vasoconstriction). Nitric oxide, in particular, is a crucial signaling molecule made in the endothelium that not only relaxes the underlying vascular smooth muscle to enhance blood flow but also possesses antiinflammatory and antithrombotic properties, inhibiting platelet aggregation and leukocyte adhesion.

When thinking of sauropods, we return to the environment in which they lived, which was quite hypoxic, and the rapid growth they had to have had in order to achieve their enormous size. One way in which the vasculature could have played a role in this was the endothelium. Hypoxia induces a process called angiogenesis, which causes expansion of endothelium to form new blood vessels. The molecular mechanisms in which this occurs are interesting, as they involve molecules that directly sense the hypoxic environment, called hypoxia-inducible factors that directly bind to DNA to turn on genes. In this way, it is possible the sauropods were pushed along, during development, by the hypoxic atmosphere acting on the endothelium.

Layered over the endothelium are smooth muscle cells, which are essential for the mechanical motion of the blood vessel, thus directly regulating blood vessel diameter and, consequently, blood flow and pressure. These cells are spindle-shaped, nonstriated, and typically arranged in layers within the walls of blood vessels, particularly in arteries and arterioles. Unlike skeletal muscle cells, smooth muscle cells operate under involuntary control, responding to a variety of stimuli such as neural signals, hormones, and local factors. One of their primary functions is to mediate vasoconstriction and vasodilation, processes critical for maintaining vascular tone and ensuring proper blood distribution. During vasoconstriction, smooth muscle cells contract, narrowing the vessel lumen, increasing vascular resistance, and elevating blood pressure. Conversely, during vasodilation, these cells relax, widening the vessel lumen, decreasing resistance, and lowering blood pressure. This dynamic regulation is vital in responding to the body's fluctuating demands, such as during physical activity or in response to changes in body temperature.

The activity of smooth muscle cells is modulated through several mechanisms involving both extrinsic and intrinsic factors. For example, sympathetic nerves innervate smooth muscles cells and induce vasoconstriction. Additionally, hormones such as angiotensin II and vasopressin can stimulate vasoconstriction. Conversely, local factors like nitric oxide and prostacyclin, produced by endothelial cells, promote smooth muscle relaxation and vasodilation. Both the constriction and dilation of blood vessels are tightly regulated by the number of calcium ions that enter the smooth muscle cells; for example, an increase in intracellular calcium concentration triggers the interaction between actin and myosin filaments, leading to contraction.

The other main cell type to consider in the vasculature is the sympathetic nerves, as they provide a direct link from the central nervous system to the blood vessels to regulate how much constriction or dilation, and thus blood flow, moves through the tissue. The sympathetic nervous system, a part of the autonomic nervous system, controls the constriction and dilation of blood vessels through the release of neurotransmitters such as norepinephrine. These neurotransmitters bind to adrenergic receptors on the smooth muscle cells. Vasoconstriction increases vascular resistance, which in turn raises blood pressure, ensuring that blood pressure is maintained at levels sufficient to meet the body's metabolic demands. This mechanism is essential for maintaining adequate blood pressure and ensuring that blood is appropriately distributed to various organs and tissues according to the body's needs.

In addition to regulating blood pressure, sympathetic innervation of the vasculature is integral to the body's response to stress, commonly known as the "fight or flight" response. During stress or physical activity, the sympathetic nervous system increases its activity, leading to widespread vasoconstriction in nonessential areas such as the skin, gastrointestinal tract, and kidneys. This conserves blood flow for vital organs and muscles that are actively engaged in the stress response or physical exertion. Concurrently, sympathetic activation causes vasodilation in skeletal muscles and the coronary arteries supplying the heart, enhancing blood flow and thus oxygen and nutrient delivery to these critical areas. This selective redistribution of blood flow allows for increased physical performance and rapid response to stressful stimuli.

Organizing the vasculature

We move from the main cell types of the vasculature to how the vascular system is organized as it receives oxygenated blood from the heart in arteries, to capillaries where the O_2 is delivered and the CO_2 taken up, and back into the veins where the deoxygenated blood is taken up. These broad

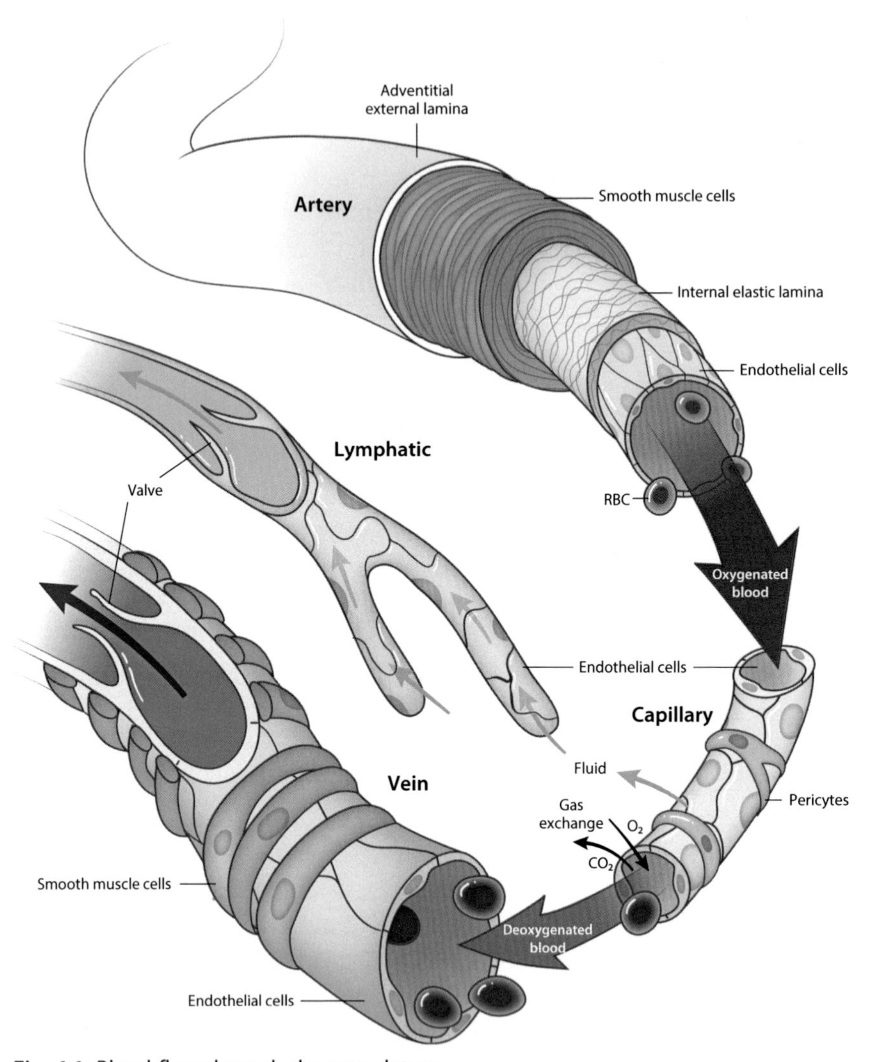

Fig. 4.1 Blood flow through the vasculature.

divisions of the vasculature are important, as they are adapted to each component of the oxygen delivery pathway (Fig. 4.1).

We begin with arteries, which are the blood vessels responsible for carrying oxygen-rich blood away from the heart to the tissues and organs of the body, with the exception of the pulmonary arteries, which carry deoxygenated blood to the lungs for oxygenation. The structure of arteries is specifically adapted to their function, featuring thick, elastic walls that can withstand and regulate the high pressure of blood being pumped from the heart. Arterial walls are composed of three layers: the tunica intima, tunica media, and tunica externa. The tunica intima is the innermost layer, consisting of a smooth layer of endothelial cells that reduces friction and provides a barrier between the blood and the vessel wall. The subendothelial layer beneath the endothelium includes a basement membrane and a connective tissue, providing structural support. The internal elastic lamina, a layer of elastic fibers, separates the tunica intima from the tunica media, allowing the artery to stretch and recoil.

The tunica media is the middle and most prominent layer of the arterial wall, consisting primarily of smooth muscle cells and elastic fibers. This layer is crucial in maintaining blood pressure and regulating blood flow through vasoconstriction and vasodilation. In large elastic arteries, such as the aorta and its major branches, the tunica media contains a high density of elastic fibers, which enables these vessels to absorb the pulsatile force of the blood ejected from the heart and to recoil, propelling blood forward. In contrast, muscular arteries, such as the coronary and renal arteries, have a tunica media dominated by smooth muscle cells, allowing for precise regulation of blood flow to various regions of the body by altering the diameter of the vessel lumen. The external elastic lamina, another layer of elastic fibers, separates the tunica media from the tunica externa. This elastic property of arteries diminishes as they branch into smaller arterioles, where the tunica media primarily consists of smooth muscle to control resistance and direct blood flow into capillary networks.

The tunica externa, or adventitia, is the outermost layer of the artery and consists mainly of collagen fibers and elastic tissue. This layer provides structural support and protection, anchoring the artery to surrounding tissues. In large arteries, the tunica externa also contains the vasa vasorum, a network of small blood vessels that supply the outer parts of the arterial wall with oxygen and nutrients. The sympathetic nerves are embedded in the tunica media. As we'll learn even more later, the arteries are especially important in the regulation of blood flow to the organs, and this is the reason they are more structurally complex and have more innervation from smooth muscle cells.

Moving from the arteries, the blood vessels continue to arborize and get smaller until they reach the capillaries. Capillaries are the smallest and most numerous blood vessels in the circulatory system, making up a huge percentile, the largest area of the vasculature. The capillaries are as small as the red blood cells of an organism—any smaller—and the red blood cells would get stuck! But that small diameter provides the most minimal distance for oxygen to move from the red blood cell to the tissue outside the capillary.

Capillaries are critical in the exchange of gases (dispensing O_2 to the tissue and taking up CO_2), nutrients, and waste products between blood and surrounding tissues. Their walls are composed of a single layer of endothelial cells, which allow for efficient diffusion of substances. This endothelial layer is supported by a thin basement membrane that provides structural integrity and selective permeability. On top of the endothelial are a cell type called pericytes, which are derived from smooth muscles, but unique in their own way. The function of these cell types is still being delineated. Most capillaries have a complete endothelial lining with tight junctions between cells, allowing only small molecules, such as water and ions, to pass through. These capillaries are found in the muscle, skin, lungs, and central nervous system, where they contribute to the blood-brain barrier. Capillaries are generally not more than ~10 μm from cells in a tissue to ensure the cells do not become hypoxic. Capillaries are literally everywhere! For a sauropod, perfusing these capillaries across their gigantic body would be paramount, and so how they did that is discussed in what follows. But first, we move from capillaries to veins.

The functional efficiency of veins extends beyond simple blood return from capillaries; they play a pivotal role in maintaining overall cardiovascular homeostasis and fluid balance. Veins are highly compliant vessels, meaning they can stretch and accommodate varying volumes of blood with little change in pressure, which makes them essential in regulating blood volume and pressure. This compliance allows veins to act as reservoirs for blood, holding ~60% to 70% of the body's total blood volume at any given time. This capability is crucial during periods of physical exertion or hemorrhage, where the venous system can mobilize blood reserves to maintain cardiac output and blood pressure. The sympathetic nervous system also influences this process by causing venoconstriction, reducing their red blood cell capacity, and increasing venous return to the heart (next chapter). This venoconstriction helps maintain perfusion to vital organs during metabolic stress or blood loss.

Furthermore, the microstructure of veins includes the venous sinuses, which are large, flattened veins with extremely thin walls, found primarily in the brain (dural venous sinuses) and the heart (coronary sinus). These venous sinuses facilitate the collection and drainage of blood from specific organs, playing a vital role in cerebral and cardiac function. The dural venous sinuses, for instance, drain deoxygenated blood and cerebrospinal fluid from the brain and direct it into the internal jugular veins. The coronary sinus collects deoxygenated blood from the myocardium and channels it into the right atrium. The interplay between the venous system and lymphatic system also highlights the importance of veins in fluid balance and immune function. Excess interstitial fluid and plasma proteins that escape from capillaries into tissues are collected by the lymphatic system and eventually returned to the venous circulation, maintaining homeostasis and preventing edema.

Unconnected from the circulatory system but equally important is the lymphatic system. The lymphatic system is responsible for maintaining fluid balance, transporting dietary lipids, and facilitating immune responses. It consists of a network of lymphatic vessels, lymph nodes, and lymphoid organs, including the spleen, thymus, and tonsils. Lymphatic vessels begin as blind-ended capillaries in the interstitial spaces of tissues, where they collect excess interstitial fluid, termed lymph, that leaks out from blood capillaries. These lymphatic capillaries have a unique structure: their endothelial cells overlap loosely to form one-way mini-valves, which open when interstitial pressure is high, allowing fluid to enter, and close when pressure is lower, preventing lymph from leaking back out. As the lymphatic capillaries converge into larger lymphatic vessels, these vessels possess valves like those in veins, ensuring unidirectional flow toward the thoracic cavity. The lymph is eventually returned to the bloodstream through the thoracic duct and the right lymphatic duct, which empty into the left and right subclavian veins, respectively. This system not only prevents the accumulation of excess interstitial fluid, thereby preventing edema, but also serves as a conduit for immune cells and other substances involved in immune surveillance and response.

Lymph nodes, strategically located along the lymphatic pathways, act as filtration and immune activation sites. The nodes are also sites of lymphocyte proliferation; they swell when fighting an infection due to the rapid multiplication of lymphocytes. Additionally, the lymphatic system plays a vital role in lipid absorption through specialized lymphatic vessels called lacteals in the villi of the small intestine. Lacteals absorb dietary lipids and lipid-soluble vitamins, forming a milky fluid called chyle, which is transported through the lymphatic system and eventually emptied into the bloodstream via the thoracic duct. This integration ensures that lipids are efficiently transported to where they are needed or stored in the body.

The dynamic movement of lymph through the lymphatic system is driven primarily by the contraction of skeletal muscles, respiratory movements, and the smooth muscle contractions in the walls of the lymphatic vessels. When skeletal muscles contract during physical activity, they compress the lymphatic vessels, pushing the lymph toward the thoracic cavity. Similarly, the pressure changes during breathing also facilitate lymph flow. Smooth muscle in the walls of larger lymphatic vessels rhythmically contracts to propel lymph forward. The lymphatic system also plays a role in immune surveillance and the immune response by facilitating the transport of antigen-presenting cells, called dendritic cells, to lymph nodes where they can interact with other important immune cells. This interaction is crucial in targeting and eliminating pathogens. The efficient functioning of the lymphatic system is thus essential for fluid balance, nutrient transport, and robust immune defense.

Compared with arteries, capillaries, and veins, lymphatics may be the more unique structure between mammals and birds/lizards. Although between all three the lymphatics serve to regulate fluid balance, the organization is slightly different. Lizards utilize lymphatic "hearts" to move their lymph primarily due to the low blood pressure in their system. As we shall see next, the assumption is that the sauropod could not have had low blood pressure in order to efficiently move blood throughout its body, so it's doubtful sauropod had similar structures. However, both birds and lizards do have less-defined lymph nodes and rely more on diffuse lymphoid tissues and aggregates scattered throughout the body.

Pressuring the resistance to flow

Next, we move to how the vasculature, in all its parts, functions. We start with blood pressure as it's critical to moving the red blood cells. Blood pressure is the force exerted by circulating blood on the walls of blood vessels and is a critical factor in driving blood flow. Blood pressure is a measurements of systolic pressure over diastolic pressure and is usually expressed in millimeters of mercury (mmHg). Systolic pressure represents the force generated by the heart during contraction, while diastolic pressure represents the pressure in the arteries when the heart is at rest between beats. The maintenance of an appropriate blood pressure gradient across the vascular system is essential to ensure that blood can flow from areas of higher pressure to areas of lower pressure, facilitating the delivery of oxygen and nutrients to tissues and the removal of metabolic waste products.

Within the vasculature, the relationship between blood flow, resistance within the blood vessel to the blood flow, and the blood pressure is a fundamental concept to understand the dynamics of blood circulation. Conceptually, this can be seen with a re-arrangement of Ohm's law in electrical circuit where blood flow (*Q*) through a vessel is directly proportional to the pressure difference (ΔP) between the two ends of the vessel and inversely proportional to the resistance (*R*) within the vessel. This is expressed as:

$$Q = \Delta P / R$$

This equation highlights that for a given pressure difference, the flow of blood increases as resistance decreases, and vice versa. This relationship is crucial in maintaining adequate tissue perfusion and ensuring that all organs receive the necessary oxygen and nutrients to function properly.

Resistance in the vasculature is influenced by several factors, including the length (*L*) and radius (*r*) of the blood vessels and the viscosity (η; thickness) of the blood. According to Poiseuille's law, resistance is directly proportional to the length of the vessel and the viscosity of the blood and inversely proportional to the fourth power of the radius of the vessel, expressed as:

$$R = 8\eta L / \pi r^4$$

According to this equation, even small changes in the radius of a blood vessel can have a significant impact on resistance. For example, a 10% decrease in the radius of a vessel can result in ~52% increase in resistance, dramatically affecting blood flow. This principle is the basis for the regulation of blood flow through vasoconstriction and vasodilation, noted earlier, where the smooth muscle in the vessel walls contracts or relaxes to decrease or increase the vessel radius, respectively.

The control of vessel diameter, and consequently resistance, is primarily achieved through the contraction and relaxation of the smooth muscle in the walls of arterioles and small arteries. These vessels, often referred to as resistance vessels, play a pivotal role in regulating blood flow to different tissues and organs. Vasoconstriction, the narrowing of blood vessels, increases resistance and decreases blood flow, while vasodilation, the widening of blood vessels, decreases resistance and increases blood flow. As noted before, this regulation is influenced by various factors including neural signals, such as sympathetic nervous activity that releases norepinephrine to cause vasoconstriction, hormones like adrenaline and angiotensin II that also promote vasoconstriction, and local factors such as nitric oxide and endothelin, which cause vasodilation and vasoconstriction, respectively. These mechanisms ensure that blood flow is adjusted according to the metabolic needs of tissues, maintaining homeostasis and responding dynamically to the body's varying demands.

Additionally, vascular resistance is not uniformly distributed throughout the circulatory system but is strategically modulated to optimize blood flow and pressure. The largest drop in blood pressure occurs across the smallest arteries, after many arborizations. These types of arteries, also called arterioles, or resistance arteries, have the greatest resistance. This allows for fine-tuned control of blood flow into the capillary beds, where nutrient and gas exchange occur. The ability to modulate resistance in specific vascular beds enables the body to prioritize blood flow to essential organs like the heart, brain, and muscles during periods of increased demand, such as exercise or stress. Conversely, during rest, blood flow can be directed to other areas such as the gastrointestinal tract for digestion. This selective distribution is critical for efficient function and survival, ensuring that critical tissues receive adequate perfusion under varying conditions.

The interplay between pressure, flow, and resistance is also influenced by the elasticity of blood vessels, particularly arteries. Elastic arteries, such as the aorta, can stretch to accommodate the surge of blood from the heart during systole and then recoil during diastole, helping to maintain continuous blood flow and smooth out the pulsatile nature of blood ejection from the heart (more on this in the next chapter). This elastic property is known as compliance, and it plays a critical role in dampening the fluctuations in blood pressure and ensuring a steady flow of blood through the vascular system. Reduced compliance can lead to increased systolic blood pressure and a greater workload on the heart.

Autoregulation is another important mechanism that governs the relationship between flow, resistance, and pressure in the vasculature. It refers to the ability of tissues to regulate their own blood flow despite changes in systemic blood pressure. This is achieved through the constriction or dilation of arterioles in response to metabolic, myogenic, and endothelial signals. For instance, during periods of increased metabolic activity, such as exercise, tissues produce metabolites that cause local vasodilation, thereby reducing resistance and increasing blood flow to meet the heightened oxygen and nutrient demands. Conversely, if blood pressure rises, myogenic responses cause arterioles to constrict, maintaining consistent blood flow and protecting the capillary beds from damage.

It might be expected that capillaries would actually have the highest resistance throughout the vasculature due to their extra-small diameter and thus limitation to blood flow. It is true that individual capillaries have high resistance. However, as noted earlier, capillaries have the largest cross-sectional area of any part of the vasculature, and due to arborization, the capillaries are arranged in parallel (rather than all lined up end-to-end). This arrangement significantly reduces the overall resistance of the capillary bed, and makes their net contribution to the overall resistance in the vasculature quite minimal.

Finally, the venous system, which returns blood to the heart, also contributes to the relationship between flow, resistance, and pressure. Veins have thinner walls and larger lumens compared with arteries, making them highly compliant and capable of holding a large volume of blood. This compliance enables veins to act as capacitance vessels, storing blood and helping to maintain venous return to the heart. Venous return is influenced by factors such as the skeletal muscle pump, respiratory pump, and venoconstriction mediated by the sympathetic nervous system. These mechanisms work together to ensure that blood is effectively returned to the heart, maintaining the pressure gradient necessary for continuous blood flow throughout the circulatory system.

The vasculature under pressure

There are three fundamental types of pressure within the vasculature the sauropod had to contend with, and nearly all of them were exaggerated due to its size—transmural, driving, and, perhaps the most important in a sauropod, the hydrostatic pressure (Fig. 4.2).

Transmural pressure refers to the pressure difference exerted across the wall of a blood vessel, which plays a key role in determining the vessel's shape and diameter under various physiological conditions. Essentially, it is the difference between the pressure inside the vessel (intraluminal pressure) and the pressure outside the vessel (extraluminal pressure), which can include the pressure exerted by the surrounding tissue depending on the location of the vessel. This pressure gradient is crucial for maintaining vascular integrity and function, influencing not only the mechanical forces acting on the vessel wall but also the vessel's ability to respond to physiological and pathological changes. For instance, in arteries, a high transmural pressure can lead to vessel stretching and, over time, to adaptive or maladaptive structural changes, such as hypertrophy of the vessel wall. Conversely, a low transmural pressure can cause vessels to collapse, particularly in veins where pressure is lower. This is vitally important in the venous system of the legs, where pressures must be regulated to assist blood flow back to the heart against gravity (more on this in what follows).

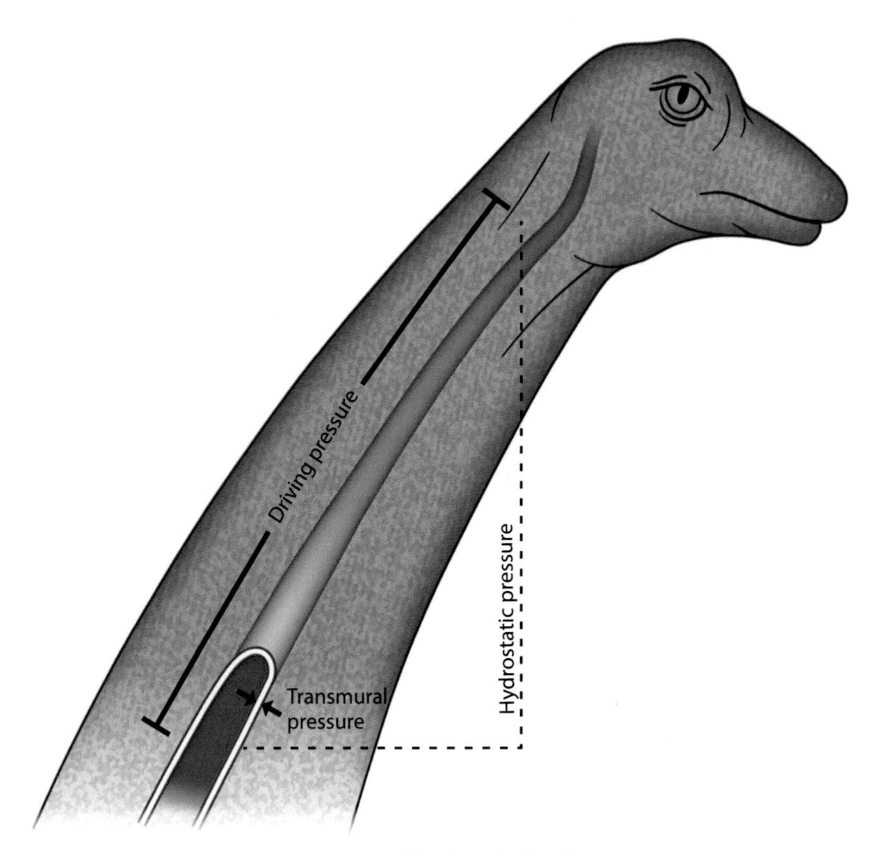

Fig. 4.2 Vascular pressure the sauropod had to deal with.

Driving pressure in the vasculature, often referred to as perfusion pressure, is a critical parameter that dictates the flow of blood through the vascular system. It is defined as the difference in pressure between two points within the vasculature, typically measured between the arterial input and the venous return. Essentially, it is the force that propels blood through the circulatory system, overcoming the vascular resistance encountered along the way. This pressure gradient is created by the pumping action of the heart, specifically the left ventricle, which generates systolic pressure that drives blood into the arterial system. The greater the driving pressure, the more effectively blood can be pushed through arteries, capillaries, and veins, ensuring adequate tissue perfusion and the delivery of oxygen and nutrients required for cellular function.

Hydrostatic pressure in the vasculature is a fundamental force influencing the movement of fluids and solutes across capillary walls, playing a pivotal role in maintaining the body's fluid balance and the effective circulation of blood. Hydrostatic pressure is the pressure exerted by the blood within the blood vessels due to the force of gravity and the pumping action of the heart. As one can imagine in the long-necked sauropod, assuming its head and neck are lifted, hydrostatic pressure in the sauropod, like the giraffe, must have been quite an important pressure to overcome. The hydrostatic pressure is the major driving force that pushes water and small solutes out of capillaries into the surrounding tissue spaces. This pressure varies significantly along the vascular tree, being highest in the arteries where blood is pumped directly from the heart and gradually decreasing as the blood moves through arterioles and capillaries, reaching its lowest points in the venous system. In the arterial end of capillaries, hydrostatic pressure is sufficiently high to overcome the opposing osmotic pressure, leading to the net filtration of fluid out of the capillary.

Conversely, the hydrostatic pressure in the veins assists in the return of blood to the heart, facilitated by the venous valves and muscle contractions, which help to counteract the effects of gravity, particularly in the limbs. The dynamic interplay between hydrostatic pressure and osmotic pressure determines the net movement of fluids across capillary membranes. At the venous end of capillaries, where hydrostatic pressure falls below the osmotic pressure exerted by plasma proteins, fluid is reabsorbed from the tissue spaces back into the bloodstream, thus conserving body fluids and maintaining circulatory volume. This balance can be disrupted by various pathological conditions, such as hypertension, which increases hydrostatic pressure and can lead to excessive fluid filtration, contributing to edema. Conversely, hypotension can reduce hydrostatic pressure too much, impeding efficient nutrient delivery and waste removal.

Gravity, the sauropods vascular foe

Remember a sauropod stood five-stories tall—that applies significant gravity. Substantial cardiovascular adjustments occur with changes in body position due to the influence of gravity on the cardiovascular system's pressures. In animals where the heart aligns horizontally with the body, such as the blue whale, the cardiovascular system only contends with pressure related to flow and resistance. The whale's heart, despite its massive size, doesn't. This is unlike the case of the sauropod, where gravity imposes additional pressure differences between the heart and body parts at different levels. This scenario applies to humans, elephants, and giraffes, with the latter having a long neck that faces considerable vertical distance from the heart, similar to what sauropods would experience.

The impact of gravity is most pronounced in the lower extremities of standing animals and even more in the case of the sauropod. For instance, in humans, intravascular pressure can rise by up to 100 mmHg due to the blood's weight in the arteries and veins of the legs. This increase doesn't affect blood flow but significantly elevates transmural pressure in veins and capillaries, leading to vein enlargement and increased capillary filtration, potentially causing edema. Without mechanisms to counteract gravity's effects, brain perfusion would diminish, leading to loss of consciousness within 10–20 min. This would pose a considerable physiological challenge for sauropods.

Vascular physiology offers solutions such as sympathetic vasoconstriction of arteries, which can reduce arterial pressure drops but doesn't sufficiently lower capillary pressure due to persistently high venous pressure. A more effective mechanism is the skeletal muscle pump, where muscle contractions compress blood vessels, expelling venous blood and lymphatic fluid from the lower limbs. Postcontraction, one-way valves in veins and lymphatic vessels prevent fluid backflow, temporarily supporting the fluid's weight. This action significantly lowers venous pressure immediately after muscle contraction, which gradually increases as veins refill. Consequently, capillary pressure and filtration rates are reduced, decreasing the lymphatic load and mitigating gravity's impact on the cardiovascular system. This brings us back to the sauropod and the amount of blood and fluid weight each of those legs must have had to carry. Conceptually, the legs of sauropods were therefore quite muscular, and the fossil record bares this out in terms of the strength and size of the sauropod's bones in the legs to support such strong muscles. So yes, those strong, thick sauropod legs were certainly for locomotion, but unquestionably, they had to be strongly utilized to move the fluid back to the heart to prevent edema (Fig. 4.3).

Because of how intricately linked the cardiovascular system is to homeostasis, the net effect of increased fluid retention (i.e., edema) in the legs causes an increase in sympathetic nerve activity and thus heart rate and cardiac output (see next chapter) to compensate for the increased fluid loss in the legs. This one of the reasons why in humans, our heart rate and blood pressure are higher during the day then when we lay down at night—our legs are not fighting as much hydrostatic pressure. Sauropods, like a giraffe, likely had very high blood pressure at baseline, making the hydrostatic pressure and edema in the legs a crucial problem.

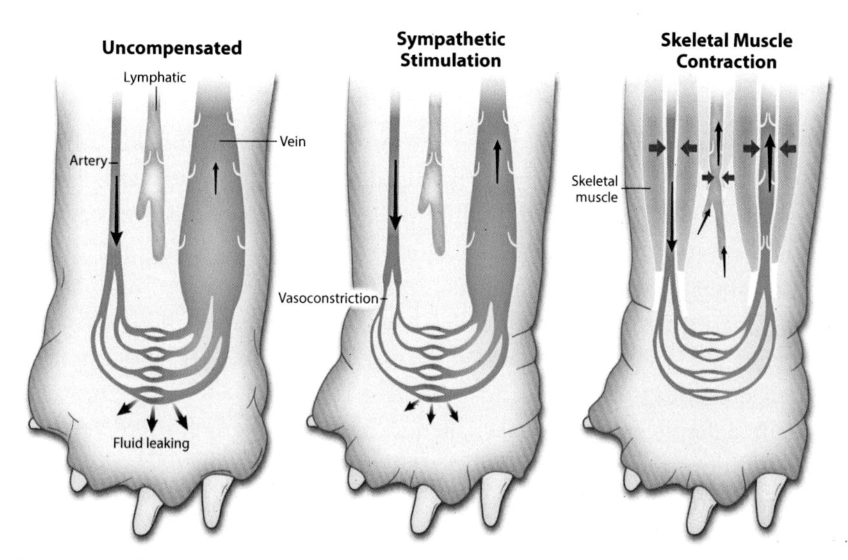

Fig. 4.3 Dealing with hydrostatic pressure in the sauropod legs (and elsewhere).

The adaptation of giraffes to its long neck involves several physiological mechanisms that protect against various challenges, including edema. With a vertical distance from the heart to the hooves of around 2 m, giraffes experience significantly higher intravascular pressures in their legs than other mammals. However, unlike a sauropod, the legs of a giraffe do not have the same skeletal muscle build (we are assuming the sauropods do based on their bone structure). Despite these pressures, edema in giraffes is prevented through a combination of structural adaptations in the large veins, conduit arteries, small arteries, and capillaries. Bicuspid valves in the femoral/tibial veins prevent capillary overload during walking or running, while morphological changes in the arteries, such as thickening of the media and reduction of lumen size, provide dynamic hemodynamic resistance. These adaptations ensure effective blood flow and prevent edema formation despite the gravitational challenges giraffes face. It is possible the sauropod also had some of these adaptions.

The brain, under pressure

An important consideration for the sauropod, assuming high blood pressure pushes enough blood upward to perfuse the brain, is how the brain is perfused. It is not likely their brains had any type of cognitive thought, but the functions of the brain still required adequate blood supply, and getting that from a distant heart presents some issues. To try to get an idea out of this, we start by looking at what the giraffe does with its brain being localized so far away from the heart.

The control of cerebral arterial pressure and flow in giraffes presents unique challenges, particularly during head movements downward, associated with drinking. Giraffes do exhibit an autoregulatory range of cerebral blood flow—like the systemic circulation—in response to changes in arterial pressure, suggesting effective mechanisms for maintaining perfusion to the brain. When giraffes lower their heads to drink, complex hemodynamic changes occur, including increased cranial arterial pressure and cessation of jugular venous flow. These adaptations ensure that cerebral perfusion is maintained despite gravitational challenges associated with head movement.

In addition, there is a unique presence in the giraffe's neck of what's called a rete mirabile, a dense and complex meshwork of small blood vessels (Fig. 4.4). These vessels are typically located at the base of the brain or along the major arteries leading to the brain. Conceptually, one of the primary functions of the rete mirabile is to facilitate countercurrent exchange. This mechanism allows for the efficient transfer of heat, gases, and other substances between adjacent blood vessels. In this arrangement, arterial and venous blood flow in opposite directions, enabling the exchange of properties like heat and oxygen between the two blood streams. Along those same

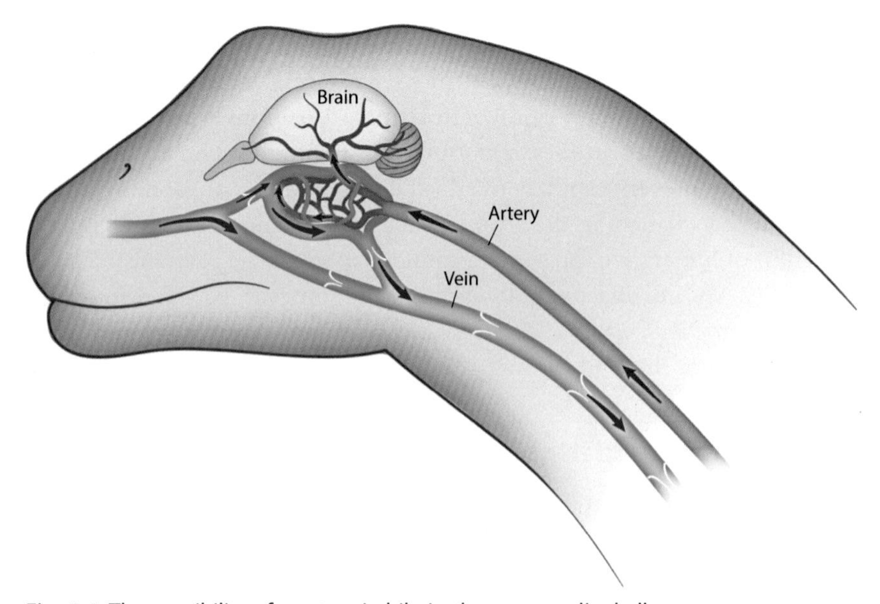

Fig. 4.4 The possibility of a rete mirabile in the sauropod's skull.

lines, it is thought the rete mirabile is important in thermoregulation, especially in species exposed to varying environmental temperatures. By adjusting the heat exchange between arterial and venous blood, the rete mirabile helps maintain a stable brain temperature. This function is crucial for proper neurological function, as the brain is sensitive to temperature fluctuations. Even more interesting in relation to the sauropod is that it seems to also be important for regulating significant changes in blood pressure, such as in animals that dive. This protective function is crucial for maintaining the integrity of delicate neural tissues, preventing damage that could result from sudden pressure fluctuations.

The structure and prominence of the rete mirabile vary among different species, reflecting adaptations to their specific environmental challenges. In fish, the rete mirabile often plays a role in regulating buoyancy and gas exchange in the swim bladder. In birds and mammals, it primarily serves thermoregulatory and oxygenation functions. For instance, in some diving mammals, the rete mirabile helps manage the distribution of oxygen during deep dives, allowing these animals to remain submerged for extended periods. In giraffes, it's hypothesized to be especially important in regulating those pressure differences when the neck bends down to drink. One can imagine the sauropod also possess a vast mete mirabile for both thermoregulation and adjustment in blood pressure. There are no indentations in sauropod's skulls to indicate the presence of such a network, but it is also unlikely they would leave such an indentation, like most soft tissue, in the first place. However, the similarity with giraffes in terms of the neck, the complex thermoregulation of the sauropod, and oxygen availability points to three significant physiological problems that could be mitigated by the presence of a rete mirabile.

We now start to tackle the huge problem of a sauropod of how the blood is moved from the heart, up the neck, and perfuse the brain in the first place. Again here we will use a giraffe as a model system for the sauropod with some of the caveats already mentioned. This will then further be discussed in the next chapter on what role the heart may play in perfusing the brain with blood.

The siphon theory of sauropod blood flow up the neck

The siphon theory of blood flow for the sauropod, particularly in relation to the giraffe, posits that a siphoning mechanism aids the circulation of blood along the extensive length of the neck, facilitating the return of blood to the heart without imposing excessive energy demands on the cardiac muscle. This theory is grounded in the principles of fluid dynamics, where gravity assists in drawing the blood down the vasculature once it reaches the upper extremities of the animal, such as the head, in a manner like water being siphoned through a hose. The heart pumps blood up the neck with considerable force, overcoming gravitational resistance, and then gravity aids its return, creating a continuous and efficient loop. Siphon works because of gravity pulling down on the taller column of liquid leaves reduced pressure at the top of the siphon, causing a differential pressure gradient due to gravity.

Detailed examination of the giraffe's cardiovascular anatomy provides some support for this theory. The giraffe's heart is robust, with a thick muscular wall, particularly in the left ventricle, which is capable of generating high blood pressure necessary to propel blood up the long neck to the brain (more on this in the next chapter). However, if the giraffe relied solely on heart's power, it would face immense energy expenditures and potentially lethal high blood pressures, especially in the brain when bending down to drink. Here, the siphon theory suggests a natural balancing act, where the descending segment of the vascular loop helps to pull blood back to the heart, easing the workload on the heart and reducing energy consumption.

However, there are anatomical and physiological complexities that challenge the siphon theory. One significant challenge is the structure of the veins, particularly the jugular veins, which must be capable of supporting the supposed siphoning effect. In a true siphon, the tube (or vessel) must be completely filled with fluid and free from air gaps to maintain the siphoning action. In biological systems, veins are not always fully filled; they are flexible, collapsible structures and not rigid tubes; thus, they cannot consistently support a perfect vacuum needed for siphoning. This flexibility is crucial for adjusting blood flow but contradicts the requirements for a functional siphon as observed in nonbiological systems.

Further complicating the application of the siphon theory to giraffes is the impact of vascular resistance and the role of valves within the veins. Blood flow resistance occurs due to friction between the blood and the walls of blood vessels, which increases with distance, potentially negating the effects of a gravity-driven siphon over the long distances involved in a giraffe's neck. Moreover, the presence of valves within the veins, which prevent backflow and ensure that blood flows toward the heart, suggests that blood return is more actively controlled rather than being a passive process reliant on gravitational siphoning alone. This is obviously an unknown in sauropod in terms of the valves in their veins, but the long-gravitational distance is even greater than in a giraffes, likely casting doubt on this mechanism of blood flow up the neck by the vasculature.

Valve hypothesis of sauropod blood flow up the neck

Another possibility is the valve hypothesis, which provides an alternative explanation to the siphon theory for managing blood flow in animals with long necks, such as giraffes and our favorite sauropod. This hypothesis focuses on the role of venous valves in facilitating the return of blood to the heart against the considerable gravitational pull resulting from the animal's long neck. Anatomically, the giraffe's venous system in the neck is equipped with numerous one-way valves that prevent backflow of blood and ensuring unidirectional return of blood to the heart. These valves play a critical role in the giraffe's ability to quickly lift or lower its head without suffering from rapid changes in blood pressure, which could otherwise lead to fainting or high intracranial pressure.

The anatomy of the giraffe's neck veins includes a series of closely spaced valves that are more numerous than those found in many other mammals. These venous valves are strategically positioned to segment the blood column along the neck, breaking it into smaller, manageable sections that reduce the hydrostatic pressure at any given point in the venous system. This segmentation effectively diminishes the weight of the blood that needs to be supported by each valve, thereby reducing the potential for venous pooling and facilitating the return of blood to the heart even when the animal bends down.

Moreover, the physiological design of these valves is tailored to support their functionality under variable pressures. When a giraffe bends its head to drink or graze, the gravitational pressure increases dramatically, which could lead to an accumulation of blood in the head and neck if not properly managed. The venous valves closes in response to any reverse blood flow, ensuring that the blood does not surge back toward the head but instead continues toward the heart.

In addition to their structural role, these valves also contribute to the overall efficiency of blood circulation by helping to create a pressure gradient. This gradient assists in the mechanical movement of blood along the neck, complemented by the muscle contractions around the veins, described above in relation to pooling in the leg, which further facilitate venous return. Every movement of the neck and surrounding muscles thus helps to "pump" the blood back to the heart, leveraging both the anatomical configuration of the valves and the physiological movements of the animal. This is supported by the fossil record where the necks from the sauropods, although not overly muscular, do provide enough structure to hold considerable skeletal muscle. Thus, examining the veins of sauropods along the neck would be an especially important area to investigate.

Other possible vascular adaptions for the sauropod

What about other vascular adaptions the sauropods may have had to ensure efficient movement of oxygenated blood throughout their body? We have discussed extensively the role of red blood cells carrying oxygen throughout the body. This is done by the large multiprotein tetramer hemoglobin jam-packed into red blood cells. We have also discussed the different red blood cell shapes and how that may contribute to changes in the amount of oxygen that can be carried throughout the vasculature. Perhaps there was a way to take advantage of the oxygen binding of hemoglobin and store it in the vasculature until it is ready to be used? This may be particularly advantageous for a sauropod that has a vast network of tissue to keep oxygenated. There is precedent for this in a slightly different globin called myoglobin. Myoglobin is a crucial protein found predominantly in the muscle tissues of vertebrates, particularly in those muscles that require a constant supply of oxygen for sustained activity, such as cardiac and skeletal muscles.

Structurally, myoglobin is a small, globular protein composed of a single polypeptide chain with 153 amino acids and a heme group at its core, which contains an iron atom capable of binding oxygen, just like hemoglobin. Functionally, in vertebrates the myoglobin serves as an oxygen storage unit, allowing muscle cells to maintain an oxygen reserve that can be quickly mobilized during periods of intense muscular activity when the demand for oxygen outpaces the supply delivered by hemoglobin in the blood. By facilitating the diffusion of oxygen from the bloodstream into the muscles and storing it until needed, myoglobin ensures that aerobic respiration can continue efficiently, thereby preventing the onset of anaerobic conditions that lead to muscle fatigue. There is evidence in humans and other mammals that myoglobin could be present in smooth muscle from large arteries and could be facilitating the same role as in skeletal muscle. It is therefore also possible, and based on the vast size likely, sauropod also had this type of adaption: an oxygen reservoir held throughout the vasculature.

The vasculature and heart acting together

We have discussed at great length the crucial role the vasculature plays in delivering the oxygenated red blood cells throughout the sauropod and the physiological adaptions the sauropod may have had to overcome extreme physiologies, especially moving blood up its long neck. From here, we move to the heart to integrate both the respiratory and vasculature portion of these discussions.

References and further reading

Curry Rogers KA, Wilson JA, editors: *The sauropods*, Berkley and Los Angeles, California, 2005, University of California Press.

Hallett M, Wedel MJ: *The sauropod dinosaurs*, Baltimore, Maryland, 2016, John Hopkins University Press.

Hill RW, Wyse GA: *Animal physiology*, ed 2, New York, New York, 1989, Harper Collins Publishers.

Klein N, Remes K, Gee CT, Sander PM: *Biology of the sauropod dinosaurs*, Indianapolis, Indiana, 2011, Indiana University Press.

Koeppen BM, Stanton BA: *Berne and Levy physiology*, ed 6, Philadelphia, Pennsylvania, 2008, Mosby Elsevier.

Mohrman DE, Heller LJ: *Cardiovascular physiology*, ed 5, New York, New York, 2003, McGraw-Hill.

Seymour RS: Cardiovascular physiology of dinosaurs, *Physiology (Bethesda)* 31:430–441, 2016.

Smerup M, Damkjær M, Brøndum E, et al.: The thick left ventricular wall of the giraffe heart normalizes wall tension, but limits stroke volume and cardiac output, *J Exp Biol* 219:457–463, 2016.

CHAPTER FIVE

At the heart of the sauropod

In the previous chapter, we discussed the siphon and valve hypothesis of blood flow up the neck. Here we start incorporating the heart, our pump for moving red blood cells through the vasculature. How was the sauropod's heart different from or the same as the species we can relate to in the modern day, what kind of adaptions may have been sauropod hearts to allow for such species radiation, and most importantly, how could we reconcile that pressure required to get blood up the neck of the sauropod? To tackle these potential adaptations, we will begin by examining some of the basics of hearts in general and layer on the possible sauropod alterations.

I heart the sauropod

One of the big questions that is likely of interest to anyone thinking about sauropods is the size of their heart. Not only is this interesting from the point of view that sauropods were enormous, but also from the perspective of the amount of power those hearts would have to generate against the force of gravity. In the imagination, the sauropod hearts must have been absolutely massive and by far the largest organ in their body. By one estimate by Seymour in *Physiology*, 2016, the sauropod hearts would have been nearly the size of their entire thoracic body cavity to accommodate the power needed to push blood all the way up the neck to perfuse the brain assuming no adaptations! This is clearly not a workable solution as they needed space for other essential organs, the amount of heat from the heart metabolism would burn up the animal, and the food that would need to be digested to feed the energy required for those hearts would not be attainable.

So let's step back for a moment and consider what we do know about heart size in other animals of similar size (blue whale) and heart to brain distance (e.g., giraffe). Indeed, contrary to common belief, the giraffe's heart size is not disproportionately large relative to its body mass. Recent studies have shown that the giraffe heart constitutes only about 0.5%–0.6% of its body mass, which is nearly identical in terms of percent of body mass to other mammals—from blue whales to dogs to mice and to humans.

Balancing a Sauropod
https://doi.org/10.1016/B978-0-12-823303-0.00005-7

Birds have a slightly large heart-to-body percentile due to their higher metabolic activity, ranging from 0.8% to 1.5%, with a hummingbird having the highest percent heart body mass at 2%. On the opposite end of the spectrum are lizards whose heart-to-body size is approximately 0.2%–0.4% of their body size, again, this time due to their lower overall metabolic activity.

So where does a sauropod fall? If we consider they fall into the category of slightly exothermic similar to lizards based on their environmental range as laid out in Chapter 2, we might consider their heart to body size on the lower end. But we likely need to push that percentile up for two reasons. The biggest reason is that sauropod hearts have the unenviable job of using their contractions to push blood against hydrostatic pressure up their enormous neck. No matter how exothermic the sauropods may have been, this was not passive and required enormous energy and, or, clever anatomical adaptions. The second reason is that sauropods were migratory and were active in herds, meaning that unlike lizards, they were many times more active, which required at least a basal level of metabolic activity.

For these reasons, we can assume that although the sauropod heart was likely large, like that of a blue whale, the heart must have been within reasonable physiological constraints, perhaps not that far off from a higher end percentile body weight of mammals, and the lower end percentile body weight seen in birds. Assuming approximately 0.7% heart mass by body weight of an *Apatosaurus* that weighed approximately 40,000 pounds (18,000 kg), the heart would be approximately 300 pounds (135 kg); an *Argentinosaurus* weighing 120,000 pounds (54,000 kg) would have a heart of approximately 850 pounds (385 kg). All of this compared with a blue whale with a 400 pounds (180 kg) heart. Therefore, there is a strong likelihood that the sauropod had a large heart, for its body size, it was not abnormal, and allowed for room in its thoracic cavity for digestive, reproductive, and other organs. The big question we look to answer then is how did the sauropod utilize this heart efficiently, and what were its adaptions, to get blood up the neck...if it even did push blood up the neck in the first place.

The heart moves the blood through the vasculature, ensuring oxygen in the blood is delivered to the tissue where it's required for metabolism, and cellular waste (for example, carbon dioxide) is shuffled out of the tissue. Blood flow through all organs is passive and occurs only because arterial pressure is kept higher than venous pressure by the beating action of the heart. In this way, the sauropod heart had to maintain enough pressure with a massively beating heart or metabolism could not occur, and very quickly, waste products would accumulate in the tissues. The right side of the heart pumps blood through the pulmonary vessels, and the left side of the heart pumps blood to all the organs, including the heart, muscles, bones, and digestive system.

But did the sauropod have a four-chamber heart? We are making this assumption, but why? In fact, most vertebrates possess hearts with two atria and one ventricle, mammals and birds have four-chambered hearts, a feature associated with endothermy—as noted in Chapter 2, the ability to maintain body temperature. Although sauropods were likely not completely endothermic, they were also not completely exothermic. So where did the sauropod fall in terms of their heart chamber number?

The division of the heart into chambers is an important organizational advantage for high metabolic functioning organisms (remember, our brain and muscles have a very high metabolic demand!). For example, most mollusks and sea slugs have single-chamber hearts to simply move a fluid around the organism to get rid of waste and deliver nutrients. Fish have two chambers that make the pumping more efficient. Most reptiles, apart from crocodiles and alligators, have three chambers. From an evolutionary perspective, independently birds and mammals evolved four-chamber hearts. Which strongly favors the hypothesis that dividing the heart creates a highly efficient mechanism to deliver oxygen in highly metabolic animals. However, we find that the exothermic animals such as crocodiles and alligators also have four chambers. Thus, there are examples across evolutionary spectrums, and endothermic versus exothermic of animals having four-chamber hearts, with the strong driving pressure for this being separation of oxygenated and deoxygenated blood to deliver oxygen and remove waste more efficiently. With the sheer size of the sauropod that had to be covered by the circulatory system, and the low-oxygen environment in which the sauropod lived, there is strong reason to suspect that the sauropod also had a four-chamber heart (Fig. 5.1).

As a tangent, the four-chamber heart is nothing compared with a cockroach that has a remarkable 13 separate chambers to continually supply fluid to their open circulatory system, enabling the bug with remarkable resiliency to damage and thrive and survive in a large range of environments—maybe that's what saves the cockroaches in a nuclear war! Thus, we may presume that the sauropod also had a four-chamber heart, but did it act in the same way birds and mammals do?

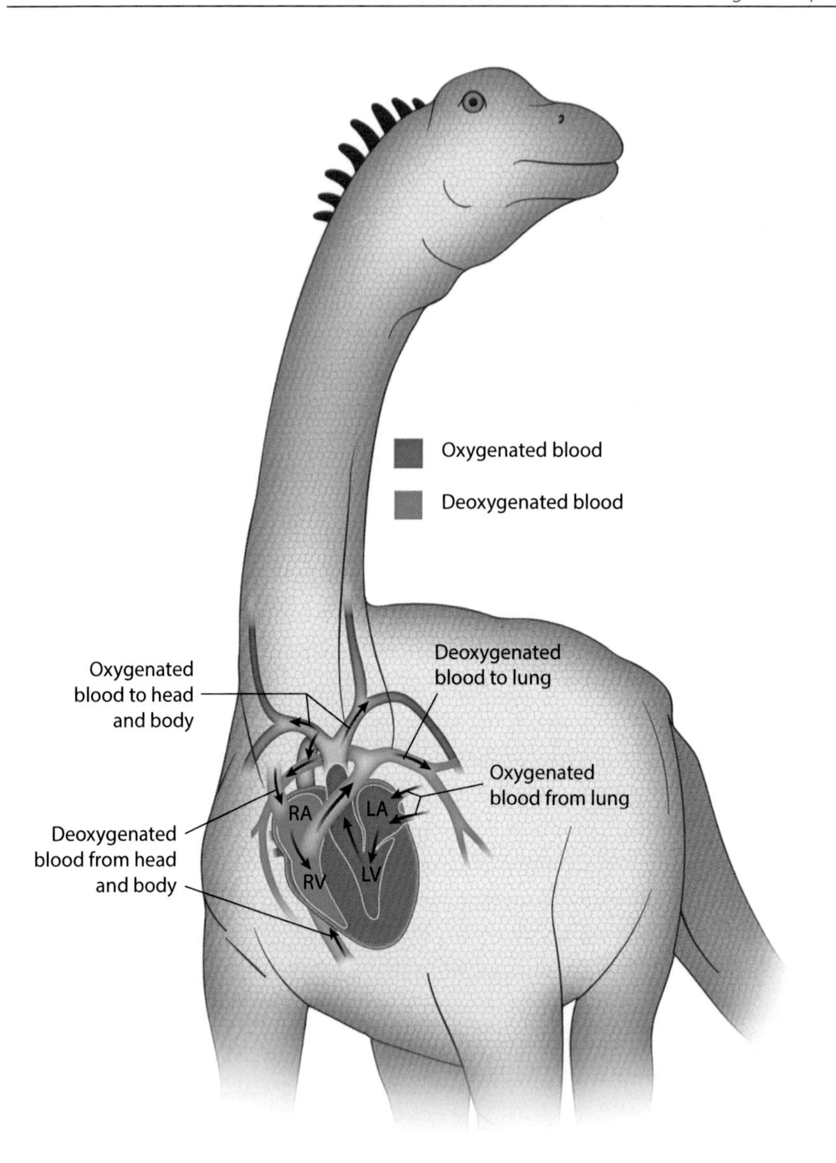

Fig. 5.1 Possible four-chamber heart of the sauropod.

The sauropod heart

We next examine the anatomical makeup of the heart and how that may alter its physiological function in a sauropod. The contractile component of the heart is provided by cardiomyocytes, which are similar to, but different from, the contractile smooth muscle cells in the vasculature, and the skeletal muscle cells in our muscle. Cardiomyocytes are the specialized muscle cells that make up the cardiac muscle, enabling the heart to contract and pump blood throughout the body. These cells are distinct from other muscle cells due to their unique structural and functional characteristics tailored for endurance and rhythmic contraction. Structurally, cardiomyocytes are cylindrical, typically branching cells that connect end to end with neighboring cells at places called intercalated discs. Across species, these discs contain cellular junctions, facilitating electrical conductivity and mechanical stability from cardiomyocyte to cardiomyocyte. This connectivity ensures synchronized contraction across the cardiac tissue, which is crucial for effective heartbeats. Additionally, cardiomyocytes contain an abundance of mitochondria to meet their high-energy demands.

Functionally, cardiomyocytes exhibit intrinsic rhythmicity and excitability, properties essential for the heart's autonomous beating. They generate and propagate action potentials through a well-coordinated process involving ion channels and the flow of ions such as sodium, potassium, and calcium. The influx of calcium ions into the cells during each action potential is particularly vital, as it triggers the sliding filament mechanism of contraction within the myofibrils. The heart's pacemaker cells, located primarily in the sinoatrial (SA) node, initiate these electrical impulses, which then travel through the atria, atrioventricular (AV) node, and ventricles, ensuring a coordinated and timed contraction of the heart chambers. From birds to mammals to reptiles, the electrical and mechanical properties of cardiomyocytes are essential for heart function.

In a four-chambered heart, like we presume was used by a sauropod, the pathway of blood through the heart moves broadly from the veins, where it is deoxygenated, into the right atrium, to the right ventricle where it is pumped to the lungs and returns with oxygenated red blood cells, to the left atrium and on to the left ventricle where it is pumped to the rest of the body. Almost all hearts have both an atrium and a ventricle, although the numbers can vary as discussed. Atria are thin-walled and have low-pressure chambers that function more as reservoirs for blood traversing to their respective ventricles (usually atrium and ventricle pair) than as an actual pump. However, the ventricles of the heart are quite muscularized as they need to pump blood either to the lungs or throughout the body. The left ventricle is especially muscularized, and the right ventricle much less so. At the heart's apex, muscle fibers twist and turn inward, forming the papillary muscles. These fibers create a robust, thick muscle around the heart's base and valves. This powerful structure not only reduces the ventricular circumference to help eject blood but also assists in valve closure.

When the cardiomyocytes in the ventricle contract, they create a circumferential tension in the ventricular walls, raising the pressure inside the left or right ventricular chamber. The pressure differences in the heart are what move blood through the heart. Once the pressure in the ventricle surpasses the pressure in the pulmonary artery (right ventricle) or the aorta (left ventricle), blood is expelled from the chamber—a process known as systole (Fig. 5.2). During systole, the higher ventricular pressure compared with the atrium keeps the valve between them closed. Conversely, when the ventricular muscle relaxes, the pressure in the ventricle drops below that of the atrium, causing the valve to open and allowing the ventricle to refill with blood, a phase called diastole. This process then starts again with systole, creating a cycle of constant blood flow through the heart and pushed through organs. This pumping mechanism is conserved across all four-chamber heart animals, including birds, mammals, and crocodiles/alligators.

Mammals and birds in particular both have an efficient heart for the complete separation of oxygenated and deoxygenated blood that is not only regulated by the chambers, but the valves control the blood moving from one area of the heart to the next. It is important to note that the blood moves from one chamber of the heart to the next through one-way valves—thus blood only goes in one direction. The valves open and close in response to the direction of pressure in the blood moving from one chamber of the heart to the next. The leaflets of the cardiac valves are composed of thin, endothelial-covered flaps firmly attached to the heart. Their movement is passive, relying on the pressure changes within the heart. This orientation

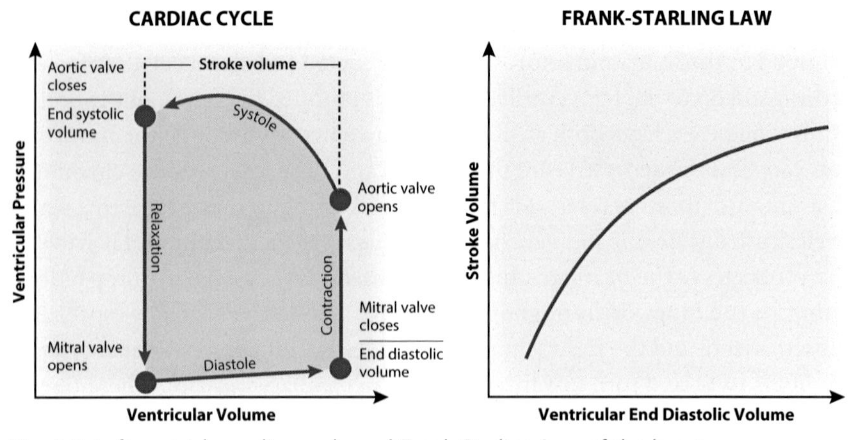

Fig. 5.2 Left ventricle cardiac cycle and Frank-Starling Law of the heart.

ensures the one-way directional flow of blood through the heart. There are two types of valves in the heart: atrioventricular and semilunar. The atrioventricular valves consist of valves separating the atrium and the ventricles of the heart. Between the right atrium and right ventricle, this valve is tricuspid, meaning it has three flaps, whereas between the left atrium and left ventricle, there only two flaps, and it is called the mitral valve. The semilunar valves connect the circulation to the heart, and both consist of three flaps; the valve between the aorta and left ventricle is called the aortic valve, and the valve between the pulmonary artery and the right ventricle is called the pulmonary valve. When blood is ejected from the ventricles, it starts to move backward, pushing the valves together to close and form a tight barrier. It is a highly efficient mechanical mechanism to ensure one-way movement of blood and segregation of oxygen and deoxygenated blood—so efficient in fact that it evolved separately in mammals and birds. Crocodiles and alligators also use valves for the one-way movement of blood through the heart.

It's interesting to note the avian ventricles are more streamlined and pointed than those of a mammalian heart, an adaptation that may aid in efficient circulation by lessening the number of fluid eddies to make the blood flow more laminar in the heart. This also occurs within the heart with both the atrial and ventricular walls in birds being notably smoother than those in other vertebrates. This smoothness extends to the structure of the heart valves. The smoother surfaces and simpler valve mechanisms reduce the friction that occurs as blood moves through the heart. Less friction means that the heart can pump blood more efficiently, with reduced workload on the cardiac muscles, which is particularly advantageous given the high cardiac output required during flight and likely high cardiac output required by a sauropod. This adaption of smooth internal ventricles is likely also something the sauropod took advantage to alter their cardiac output (see the following).

As noted earlier, alligators and crocodiles also have a four-chamber heart, but it is quite specialized to meet their physiological needs. Their atrioventricular and semilunar valves are the same ensuring separation of oxygenated and deoxygenated blood, but they also possess a small valve-like structure called the Foramen of Panizza that connects the pulmonary artery (deoxygenated blood) to the aorta (oxygenated blood), allowing for some mixing of oxygenate and deoxygenated blood. They also evolved an interesting cogwheel valve at the base of the pulmonary artery that can constrict, preventing blood flow to the lungs and redirecting to the rest of the circulation allowing for increased blood flow—the downside is decreased overall oxygen, but it does, by virtue of removing a large circulatory tract, increase the flow of the remaining oxygenated blood. These kinds of small anatomical adaptions provide a functional benefit to the alligators and crocodiles as it is thought it allows more oxygen when diving. One can imagine similar small anatomical adaptions in the sauropod heart that would provide more oxygenated blood for a short amount of time when migrating.

Blood pressure and the sauropod

An important physiology that links the last two chapters and this chapter together is blood pressure, and we have learned about several components to it, including how the vasculature works to constrict and dilate and how that changes the resistance. We now merge that with what we will learn further about cardiac output to determine blood pressure. Blood pressure is the force exerted by circulating blood on the walls of blood vessels, a fundamental aspect of cardiovascular physiology. The blood pressure of any animal with a closed circulatory system is generated by the heart as it pumps blood into the arteries and is influenced by the dynamic interplay between the heart's output and the resistance of the blood vessels. This can be represented mathematically by the following equation where CO is the cardiac output and TPR is the total peripheral resistance:

$$\mathrm{BP} = \mathrm{CO} \times \mathrm{TPR}$$

What that equation tells us is that both the cardiac output and total peripheral resistance from blood vessels work together to determine the overall blood pressure. If there was a low cardiac output, but normal blood pressure, then an increase in total peripheral resistance, e.g., increased sympathetic constriction of arteries would have to occur. In fact, the pressure is highest in the arteries due to the direct pumping action of the heart and gradually decreases as blood moves through the arterioles, capillaries, venules, and veins. This gradient ensures that blood flows in a unidirectional manner, supplying oxygen and nutrients to tissues while removing metabolic waste products. Blood pressure is a crucial driver for perfusion, the process by which tissues receive an adequate blood supply to meet their metabolic needs.

Blood pressure is also intricately linked to the physical properties of the blood and the blood vessels. The viscosity of blood, which can be influenced by factors such as hematocrit levels and plasma composition, affects the resistance to flow within the vessels. The diameter and elasticity of the blood vessels are equally important; wider and more elastic vessels can accommodate more blood with less resistance, leading to lower blood pressure. Conversely, more narrow or stiffer vessels increase resistance, thereby raising blood pressure. Remember that the endothelium from the previous chapter, which lines the inside of blood vessels, plays a vital role in modulating blood pressure by releasing vasoactive substances such as nitric oxide, which can cause vasodilation or vasoconstriction, respectively. These complex interactions ensure that blood pressure is finely tuned to meet the varying demands of the body's tissues and organs, facilitating efficient circulation and overall homeostasis.

The heart's cycle of contraction (systole) and relaxation (diastole) creates the pulsatile nature of blood pressure. During systole, the ventricles contract, ejecting blood into the aorta and pulmonary arteries, which causes a peak in pressure known as systolic blood pressure. This phase ensures that blood is pushed through the arteries, overcoming the vascular resistance encountered in the systemic and pulmonary circuits. During diastole, the ventricles relax and fill with blood from the atria, and the pressure in the arteries falls to a minimum level known as diastolic blood pressure. In addition, the elasticity of the arterial walls plays a crucial role in maintaining blood pressure during diastole, as they stretch to accommodate the surge of blood during systole and then recoil to keep the blood moving forward during the resting phase.

So what kind of blood pressure did the sauropod have? We start again by comparing with other animals and remarkably, similar to heart size, blood pressure is tightly conserved across mammalian species, with blood pressures at approximately 100–120 mmHg mean arterial pressure. The only exception among the mammals is the giraffe, which is drastically increased at approximately 250 mmHg mean arterial pressure, even though their heart, as noted earlier, is the same body weight percentile as the other species. This blood pressure is enormous and is the largest recorded of other animal species. For example, crocodiles and alligators, as well as reptiles, have blood pressure of approximately 40 mmHg due to their low metabolic activity and minimal movements, whereas birds have a blood pressure of approximately 150 mmHg mean arterial pressures. So what makes the giraffe different? The difference is the distance from the heart to the brain across that exceptionally long neck that no other species must manage. Looking at the sauropod, we see an identical comparison, regardless that their size more closely matches a blue whale. The blue whale never had to lift its neck and fight hydrostatic pressure though. Even if there was a degree of exothermic nature to their thermoregulation (e.g., the blood pressure of lizards), they would still need that higher pressure to keep blood moving. To try to dissect this further, let's launch into what cardiac output is and see if we can learn about how the sauropod's four-chamber heart may have worked with respect to blood pressure.

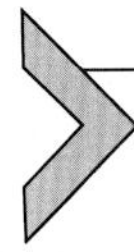

The cardiac output from the sauropod heart, part 1—Heart rate

From here we move on to how the heart delivers the blood to the rest of the body, and we will focus on mechanisms conserved across species and highly likely to have taken place in the sauropod heart. We start with possibly the most important component to heart function, which is called the cardiac output, or the amount of blood from each ventricle pumped per minute. The cardiac output reflects the efficiency with which the heart supplies oxygenated blood to the body's tissues and organs. Adequate cardiac output ensures that vital organs receive sufficient oxygen and nutrients to function properly, maintain metabolic processes, and remove waste products. The cardiac output depends on the volume of blood ejected per beat (called the stroke volume) and the number of heart beats per minute. Essentially, all influences on cardiac output must act by changing either the animal's heart rate or the animal's stroke volume. Cardiac output is one of the most important cardiovascular variables that is constantly being adjusted to meet the animals' moment-to-moment metabolic needs. For example, going from rest to strenuous exercise, the cardiac output can more than triple to provide skeletal muscle with additional nutritional supply to sustain the muscle metabolism.

We will begin by examining how an animal's heart rate is regulated. Normal rhythmic contractions of the heart occur due to spontaneous electrical pacemaker activity of cells in an area of the heart called the sinoatrial (SA) node, which sits at the far anterior end, above the atria. The signal passes from the SA node and atria to the arterioventricular (AV) node over the ventricles, causing the signal to induce a ventricular contraction. This constant movement of electrical signal from SA node and atria to AV node and ventricles coordinates heart contraction and then a reset to perform the mechanical action again. The interval between heart beats is determined by how long it takes these pacemaker cells to spontaneously depolarize.

But what does depolarization of a cell mean? To answer that question, we first must understand what membrane potential is. It is one of the most fundamental aspects of how cells regulate homeostasis and respond to

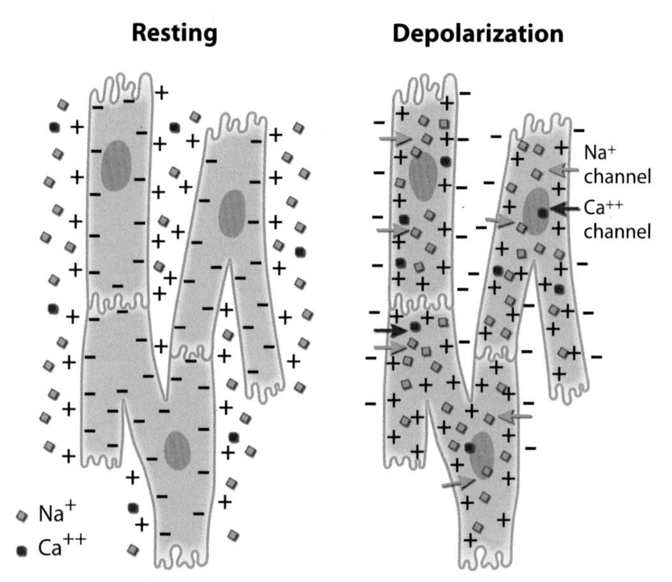

Fig. 5.3 Maintenance of cellular membrane potential.

hormones or neurotransmitters. Membrane potential refers to the difference in electrical charge between the inside and the outside of a cell membrane. This potential arises from the separation of positive and negative charges across the plasma membrane (Fig. 5.3).

In almost all cells in the body, the inside of the cell is negatively charged compared with the outside. This difference in charge is maintained by the selective permeability of the cell membrane to ions such as sodium (Na^+), potassium (K^+), chloride (Cl^-), and calcium (Ca^{2+}), as well as by the action of ion pumps and channels situated on the plasma membrane. The resting membrane potential of a cell typically ranges from −40 to −90 mV. Changes in membrane potential cause a depolarization (an increase in membrane potential as Na^+ enters the cell and switches the outside charge on the plasma membrane from positive to negative) or a hyperpolarization (a decrease in membrane potential due to potassium entering the cell). After hyperpolarization, a sodium/potassium transporter restores the membrane potential to more positive on the outside of the cell by pumping the remaining Na^+ out of the cell. It's these alterations in membrane potential that determine the contraction of a cardiomyocytes (hyperpolarization), as well as a host of other physiological changes in cells, inducing release of substances in the extracellular environment, and turning on or off other ion channels in the plasma membrane of the cell.

The autonomic nervous system plays a pivotal role in regulating the depolarization of the SA node, a critical component of heart rate control. Both the parasympathetic and sympathetic divisions of the autonomic nervous system exert direct influence on the SA node's activity. Activation of the parasympathetic nervous system leads to a decrease in SA node firing frequency and subsequently a reduction in heart rate. This effect is mediated by the release of acetylcholine by parasympathetic nerve fibers onto SA node neurons. Acetylcholine induces hyperpolarization of these cells, slowing their spontaneous depolarization. Additionally, acetylcholine diminishes the speed of the electrical signal propagation from the SA to the AV node.

Conversely, activation of the sympathetic nervous system results in the release of norepinephrine onto SA node cells. Norepinephrine promotes an increase in the depolarization interval of SA node cells, enhancing the electrical signal transmission from the SA to the AV node. Thus, sympathetic stimulation leads to an elevation in heart rate, facilitating the body's physiological response to various stimuli or stressors.

However, the regulation of heart rate is not solely determined by pacemaker (SA and AV nodes) activity on cardiomyocytes. Numerous components, including hormonal signals, neural inputs, and environmental factors, can modulate cardiomyocytes depolarization process, altering an animal's heart rate. Some of these factors include temperature and atrial wall stretch if there is an abundance of blood entering the heart. Interestingly, an abnormally high concentration of calcium in the blood or extracellular fluid can decrease the heart rate by altering the way in which cells maintain their membrane potential, making it more difficult for the cell to hyperpolarize. This bring us to an odd behavioral adaption that our long-neck friend the giraffe likes to perform.

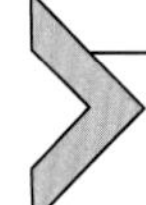

The cardiac output from the sauropod heart, part 2—Stroke volume

Besides heart rate, the other key component to regulating the cardiac output is the stroke volume, which is the volume of blood ejected per beat of the heart. In sauropods, the stroke volume would have had to be enormous to push the volume of blood needed in the sauropod circulation. There are several factors that go into stroke volume that are interrelated, allowing for the sauropod to perhaps have had adaptions that would allow the heart to function efficiently in such a large animal.

Stroke volume is a critical parameter in cardiovascular physiology, regardless of the animal species, and would be highly likely to have influenced sauropod heart function. The stroke volume is the volume of blood ejected by each ventricle during a single contraction. The physiology of stroke volume is influenced by three primary factors: *preload*, *afterload*, and *contractility*.

Preload refers to the initial stretching of the cardiomyocytes before contraction, which is largely determined by the volume of blood returning to the heart (venous return) and filling the ventricles during diastole. The degree of myocardial stretch before contraction is directly proportional to the end-diastolic volume, the volume of blood in the ventricles at the end of diastole. This relationship is described by the Frank-Starling law of the heart, which states that an increase in end-diastolic volume leads to a more forceful contraction and, consequently, a greater stroke volume (returning to Fig. 5.2). This intrinsic mechanism allows the heart to adapt to varying volumes of venous return, ensuring that the amount of blood ejected during systole matches the amount of blood returned to the heart, maintaining a balanced circulation.

Several factors influence preload, including blood volume, venous tone, and the duration of diastole. Blood volume is a primary determinant with an increase in blood volume, as seen with fluid retention, raising preload by increasing venous return. Venous tone, controlled by the autonomic nervous system, also has an important role. Sympathetic stimulation causes venous constriction, which enhances venous return and preload, while parasympathetic activity leads to venous dilation, reducing preload. Additionally, the duration of diastole, which is the filling phase of the cardiac cycle, affects preload. A longer diastole, as seen with slower heart rates, allows more time for ventricular filling, increasing end-diastolic volume and preload. Conversely, a shorter diastole, as occurs with increased heart rates, reduces ventricular filling time and preload. Based on these concepts, we would hypothesize that a sauropod would want to increase its end-diastolic volume to get the blood from the heart up the neck and to the brain, and thus, their heart rate may have been slower than we might expect.

Preload's impact on cardiac function is significant and multifaceted. By influencing the force of ventricular contraction through the Frank-Starling mechanism, preload is clearly important for regulating stroke volume and cardiac output. In heart failure, where the heart's pumping ability is compromised, optimizing preload can help enhance cardiac output. However, excessively high preload can lead to increased ventricular wall stress and exacerbation of heart failure symptoms, such as pulmonary congestion and edema. Conversely, insufficient preload, as seen in conditions with low blood volume (also called hypovolemia) or severe dehydration, can lead to reduced cardiac output and inadequate tissue perfusion.

The next stroke volume indicators is afterload, which refers to the resistance that the ventricles must overcome to eject blood during systole. It is primarily influenced by the systemic vascular resistance (blood vessels; see previous chapter) and the condition of the aortic and pulmonary valves that enter the heart. Systemic vascular resistance, also known as total peripheral resistance, is determined by the diameter of the systemic arteries and arterioles and has already been discussed. When these blood vessels constrict, systemic vascular resistance increases (the analogy of a kinked hose raising pressure elsewhere), raising the afterload. Conversely, when blood vessels dilate, systemic vascular resistance decreases, thus lowering the afterload. The condition of the aortic valve also plays a significant role in afterload. Anything condition that alters their opening alters the afterload. For example, aortic stenosis, where the aortic valve is narrowed, increases afterload because the left ventricle must generate higher pressure to force blood through the restricted valve opening.

The impact of afterload on cardiac function is multifaceted. An increased afterload requires the ventricles to generate higher pressure to eject blood, which in turn increases the myocardial oxygen demand and workload. The left ventricle, being the main chamber responsible for pumping blood into the systemic circulation, is particularly affected. To compensate for increased afterload, the ventricular myocardium undergoes hypertrophy, particularly concentric hypertrophy, where the walls of the heart thicken without an accompanying increase in chamber size. This adaptation initially helps to maintain cardiac output despite elevated afterload. However, prolonged exposure to high afterload can lead to pathological changes, including left ventricular hypertrophy, which is associated with reduced compliance, diastolic dysfunction, and eventually heart failure. This progressive remodeling impairs the heart's ability to relax and fill properly during diastole, further compromising cardiac output and leading to clinical symptoms of heart failure. As we will see further, the effects of heart failure we just described mimic a giraffe and may have mimicked what the sauropod had adapted their own heart. Why might this be the case?

According to the inverse relationship between afterload and stroke volume, as afterload increases, stroke volume decreases if other factors remain constant. This is explained by the force-velocity relationship of the cardiac muscle, which states that the velocity of myocardial fiber shortening decreases as the afterload increases. Therefore, higher afterload reduces the efficiency of ventricular ejection in the heart. For humans, this relationship is observed in conditions such as hypertension and aortic stenosis noted earlier, where elevated afterload leads to reduced stroke volume and compromised cardiac output. Predictably then, therapeutic strategies to manage high afterload include the use of drugs that induce vasodilation, reducing systemic vascular resistance and thus afterload, improving stroke volume and overall cardiac function.

The last factor influencing stroke volume is contractility of the heart. Contractility of the heart, also known as inotropy, is the intrinsic ability of cardiomyocytes to contract independently of preload and afterload. Increased contractility enhances stroke volume by enabling the heart to pump more effectively, even without changes in preload or afterload. At the cellular level, contractility is primarily regulated by the availability of calcium ions within the cardiomyocytes. During each cardiac cycle, calcium influx through voltage-gated L-type calcium channels triggers the release of more calcium from within the cells. This calcium binds to troponin, initiating the cross-bridge cycling of actin and myosin filaments that generates the contractile force of the heart muscle.

Contractility is influenced by the autonomic nervous system and circulating hormones. Sympathetic stimulation increases contractility through the release of norepinephrine and the activation of beta-adrenergic receptors on cardiac myocytes. This leads to increased intracellular calcium levels in cardiomyocytes, enhancing the force of myocardial contraction and thereby increasing stroke volume. Additionally, the hormone epinephrine, released from the adrenal medulla during stress or exercise, similarly boosts contractility of the cardiomyocytes. Conversely, parasympathetic stimulation, mediated by the vagus nerve and the neurotransmitter acetylcholine, reduces heart rate and contractility. The balance between sympathetic and parasympathetic influences on contractility is thus an important mechanism for maintaining optimal stroke volume and cardiac output under varying physiological conditions, but is especially pronounced in exercise.

The relationship between contractility and stroke volume is especially important with increased physiological demand, such as exercise or stress, where enhanced contractility ensures sufficient cardiac output to meet the body's increased metabolic needs. Again we come back to the fact that sauropods were not stationary animals, and their sheer size would necessitate changes in contractility as they walked for any amount of distance. This would be further enhanced when the sauropod had to migrate for food, stretch their neck for a tasty leaf. This would also be increased if the sauropods were swimming in water, requiring a bigger change in heart contractility for the increase in blood flow due to the increase in oxygen demand for the increased number of muscles needed as compared with walking. For this reason, the contractility component of stroke volume was likely a large determinant of the speed at which sauropods could move. If the sauropods were swimming, this may have greatly increased the contractility relationship in the heart, making it a likely less ideal method of movement for such a large animal as a sauropod.

Sauropod tension in the heart

Related to, but separate from, the components of cardiac output is wall tension in the heart. This important physiological indicator refers to the force exerted on the walls of the heart's chambers, especially the ventricles, during the cardiac cycle. It is a critical factor influencing how the heart functions to pump blood effectively. Wall tension is primarily governed by the Law of Laplace, which states that the tension in the wall of a spherical structure (such as the heart) is directly proportional to the internal pressure and radius of the chamber and inversely proportional to the wall thickness. Mathematically, it is expressed as:

$$T = (P \times R)/(2 \times h)$$

where T is the wall tension in the heart, P is the internal pressure in the heart, R is the radius of the ventricular chamber, and h is the thickness of the ventricular wall of the heart. During systole, when the ventricles contract, the internal pressure rises to propel blood into the aorta and pulmonary artery, increasing the heart wall tension. Conversely, during diastole, when the heart relaxes and fills with blood, the pressure and wall tension decrease. The heart muscle (myocardium) adapts to these changes in wall tension through hypertrophy, where an increase in wall thickness can reduce the tension experienced by the myocardial fibers. This adaptation is crucial in conditions of chronic high pressure, such as hypertension, where sustained high wall tension can lead to pathological remodeling of the heart muscle.

Putting the cardiovascular system together

An interesting correlation between pathological adaptions in human hearts and a normal healthy giraffe heart is seen when examining heart failure in humans. This likely also extends to sauropods, so let's take a moment to discuss the observations. Heart failure, where the heart cannot pump enough blood to meet the body's needs, is a leading cause of hospitalization, especially among elderly humans. Heart failure can be classified based on left ventricular ejection fraction. Ejection fraction is a measurement that quantifies the percentage of blood ejected from the heart's ventricles with each contraction. It is primarily used to assess the heart's pumping efficiency, especially that of the left ventricle. Ejection fraction is expressed as a percentage, representing the ratio of the stroke volume to the end-diastolic volume. An ejection fraction within the range of 50%–70% is considered normal, indicating that the heart is pumping enough blood with each beat. Values below this range can indicate various degrees of heart failure.

There is heart failure with preserved ejection fraction (HFpEF) or reduced ejection fraction (HFrEF) or two of the main types of heart failure we will focus on. While treatments for HFrEF have made progress, managing HFpEF remains exceptionally challenging. Interestingly, studying sauropods, and giraffes, might offer clues on unique ways to treat the disease. The reason for this possible insight is that pathological changes in the human heart's structure during HFpEF, especially ventricular hypertrophy, mirror the giraffe's heart normal structure. In both humans with HFpEF and giraffes, there is chronic high blood pressure. The giraffe's hypertension is life long, is even greater than most of the pathological human hypertension (e.g., normal human mean arterial pressure is approximately 120 mmHg, a chronic hypertensive human is approximately 150 mmHg, compared with the healthy giraffe at approximately 250 mmHg). However, unlike humans, giraffes do not experience adverse effects on exercise capacity or cardiac function that humans with HFpEF have. This resistance to HFpEF-like pathology in giraffes is crucial for their survival from predators, maintaining maximal exercise capacity for escape. In humans, left ventricular hypertrophy due to HFpEF and high blood pressure leads to extreme fibrosis, where noncellular proteins such as collagen fill the gaps between stretched cardiac myocytes. This fibrosis impairs the heart's ability to pump blood effectively, eventually leading to heart failure (Fig. 5.4). So how does the giraffe maintain such a hypertrophied left ventricle for normal function, whereas a human's hypertrophied left ventricle in HFpEF is mortal diagnosis?

Recently, genetic and molecular analyses suggest that giraffes may have evolved mechanisms to suppress cardiac fibrosis that accompany the HFpEF in humans. The fibrosis, with its accumulation of proteins outside of the cardiomyocytes, of the hypertrophied heart causes it to become stiff, and unable to relax in humans—whereas the giraffe heart never encounters this problem. It is likely the giraffe evolved mutations in specific genes, including fibroblast growth factor receptor-like 1. These receptors bind fibroblast

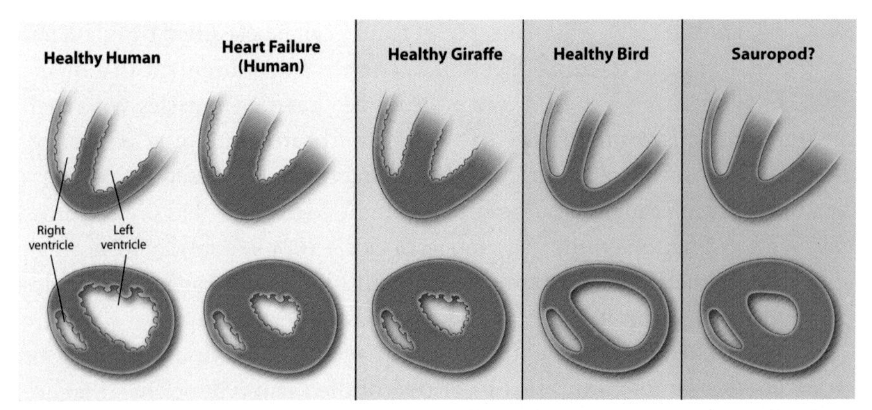

Fig. 5.4 Transverse and top view of heart structure in multiple species. *(Modified from Natterson-Horowitz B, Baccouche BM, Mary J, Shivkumar T, Bertelsen MF, Aalkjær C, Smerup MH, Ajiiola OA, Hadaya J, Wang T: Did giraffe cardiovascular evolution solve the problem of heart failure with preserved ejection fraction, Evol Med Public Health 9:248–255, 2021.)*

growth factors (FGFs), a family of proteins that bind to proteins outside of cardiomyocytes, such as heparin and heparan sulfate. Consequently, some FGFs become sequestered outside of the cardiomyocytes, within what's called the extracellular matrix and accumulate following injury or tissue remodeling, like the excessive hypertension in humans. Thus, lacking receptors for FGF could protect the heart, allowing it to stretch under high blood pressure without developing fibrosis. In this way, the giraffe has all the benefits of the pathological adaptions to high blood pressure seen in humans, but retains the ability of the heart to relax without the fibrosis. In this way, giraffes get all the benefits of the muscularized, hypertrophied heart, but don't suffer the pathologies that would normally accompany that adaption in humans.

With these adaptions in ventricular anatomy, characterized by a small left ventricular cavity, small ventricular radius, and thick ventricular walls, what happens to the cardiac output in a giraffe? In effect, those thick ventricular walls serve to normalize the overall wall tension in the heart due to the high blood pressure, and the net effect is a fairly normal stroke volume, cardiac output, and ejection fraction, in line with other animals. As we learned earlier, heart rate is also a component to cardiac output, and the heart of giraffes has also been documented to be within the normal-to-high range of mammals. But certainly there would need to be moments when they would want to increase their heart rate even with their high blood pressure (e.g., escaping predators)? Some clues may be found in some giraffe behaviors.

Weirdly, giraffes eat bones—it's not the first thing that comes to mind when you think of a giraffe because as we all know, they are herbivores. However, it has been well documented that giraffes will scavenge bones from carrion eaten by other animals. Why might giraffes do this? One clear reason is likely the need for calcium and other minerals in an environment where some of the nutritional needs are not always met with low-metabolic quality leaves and twigs. Giraffes also have long skinny legs with much less skeletal muscle than a sauropod may have had, so keeping those bones healthy is important.

But there may be another, just as important reason: to regulate heart rate. Excess calcium ions influx enhances the depolarization rate at the SA node (described earlier) by altering the membrane potential of the cells. The net effect of this is a faster generation of action potentials and thus overall an increased heart rate. Remember that cardiac output is distinctly regulated by heart rate and stroke volume. Thus, an increase in heart rate by eating bones could be a means by which giraffes could increase cardiac output. In a giraffe, this could ease some of the burden on the stroke volume after excessive running or escaping a predator (discussed further). Although sauropods were of course gigantic, there is evidence from tandem sauropod tracks that they could move rather quickly (it may be a stretch of our wording to say "running"). One could also extrapolate to sauropods that a little gnosh on some bones after a long migration or a fight with a predator could have been just enough of an extra little push to keep their heart pumping those red blood cells and ease some of the burden on the stroke volume to ensure a constant cardiac output.

Thus, if we look back on what determines blood pressure, both the cardiac output and total peripheral resistance, we find the cardiac output part of the equation is on the normal-to-low side. Thus, in the face of the well-documented high blood pressure in giraffes, the total peripheral resistance really is the main driver of the giraffe hypertension. And as described in Chapter 4, the giraffe and sauropod are both fighting immense hydrostatic pressures due to the effects of gravity. It is therefore extremely important that the hearts of a giraffe, and especially the sauropod, mitigate this with low-to-normal cardiac output. Indeed, one could argue that it would be dangerous to lower the cardiac output part of this equation too much for fear of being unable to perfuse the circulation adequately. Thus, keeping the cardiac output as normal to low-normal as possible is the best the giraffe or sauropod can do to counteract the large total peripheral resistance driving the high blood pressure. The adaption of the giraffe preserves the integrity of the heart in the face of a lifetime of high blood pressure, and almost certainly, sauropods would have had to have had similar adaptions.

Other possible heart adaptations

Throughout this chapter, we have made the singular assumption that sauropods had one, four-chamber heart. We have used the giraffe as an example of an animal sharing the long neck with the exaggerated distance between the heart and the brain, even though similar-sized organisms such as a blue whale existed. We did this because giraffes and sauropods would share the same major issue of immense hydrostatic pressure on the vasculature, and thus blood pressure, whereas the blue whale had no hydrostatic pressure to fight against in homeostatic conditions. For this reason, we may make a tentative conclusion that the giraffe and sauropod were similar in the cardiovascular system, at least in the heart. There are some possible differences that could alter this conclusion, which we will discuss in the final chapter.

Before we move on, there are a few minor issues to consider when thinking of the sauropod heart. One hypothesis that had been considered is if sauropod had multiple hearts. Another possibility is that sauropods had multiple chambered hearts. The advantage for both situations is that the replicates would help dampen the inevitably high blood pressure, putting less pressure on the heart. There are exceptionally few examples we can point to of organisms processing multiple hearts outside of the multichambered cockroach. In the cockroach's case, their circulatory system is open and those multiple chambers ensure a constant flow. An open vascular system in the sauropod would not be possible due to size and complexity. The multiple hearts or minor hearts leading up the neck is interesting, but would require a large investment of metabolic energy the sauropod could surely not achieve.

References and further reading

Aalkjær C, Wang T: The remarkable cardiovascular system of giraffes, *Annu Rev Physiol* 83:1–15, 2020.

Curry Rogers KA, Wilson JA, editors: *The sauropods*, Berkley and Los Angeles, California, 2005, University of California Press.

Hallett M, Wedel MJ: *The sauropod dinosaurs*, Baltimore, Maryland, 2016, John Hopkins University Press.

Hill RW, Wyse GA: *Animal physiology*, ed 2, New York, New York, 1989, Harper Collins Publishers.

Klein N, Remes K, Gee CT, Sander PM, editors: *Biology of the sauropod dinosaurs*, Indianapolis, Indiana, 2011, Indiana University Press.
Koeppen BM, Stanton BA: *Berne and Levy physiology*, ed 6, Philadelphia, Pennsylvania, 2008, Mosby Elsevier.
Levitzky MG: *Pulmonary physiology*, ed 6, New York, New York, 2003, McGraw-Hill.
Liu C, Gao J, Cui X, et al.: A towering genome: experimentally validated adaptations to high blood pressure and extreme stature in the giraffe, *Sci Adv* 7:eabe9459, 2021.
Mohrman DE, Heller LJ: *Cardiovascular physiology*, ed 5, New York, New York, 2003, McGraw-Hill.
Natterson-Horowitz B, Baccouche BM, Mary J, et al.: Did giraffe cardiovascular evolution solve the problem of heart failure with preserved ejection fraction, *Evol Med Public Health* 9:248–255, 2021.
Seymour RS: Cardiovascular physiology of dinosaurs, *Physiology (Bethesda)* 31:430–441, 2016.
Smerup M, Damkjær M, Brøndum E, et al.: The thick left ventricular wall of the giraffe heart normalizes wall tension, but limits stroke volume and cardiac output, *J Exp Biol* 219:457–463, 2016.

CHAPTER SIX

Ideas on sauropod kidneys and digestion

In the previous three chapters, we focused extensively on the respiratory-cardiovascular system and how its different adaptions may apply to a sauropod. Whereas, the respiratory system has two broad distinctions in land-dwelling animals between avian-type lungs and mammalian-type lungs. Next, we moved to the cardiovascular system with the conserved blood vessels and heart being more uniform and conserved across species. The two remaining areas of interest in sauropod physiology that are essential to its homeostasis are kidney function and metabolism. The kidney in particular has broad adaptions to the environment, whereas metabolism is more conserved in terms of adaptions. Both of these systems follow directly from the respiratory and cardiovascular systems. We will begin with the kidney and from there tackle metabolism in the sauropod.

The kidneys are intricate and essential organs responsible for maintaining various physiological balances within the body. In most animals, the kidneys are on either side of the spine in the abdominal cavity. They play several critical roles, including filtering blood to remove waste, balancing electrolytes, regulating blood pressure, and producing hormones.

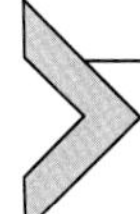

The kidneys part 1: Filtration and water regulation the nephron

Although the kidney is a remarkably multifaceted organ, we begin with the function of filtration and water regulation as this is one of the primary jobs of the kidney. We also begin with mammalian kidneys as they are the most complex, and from here we can work backward. We do this because the kidney appears to be similar embryonically across species, but depending on the complexity of the organism, the kidney increases in tissues and layers.

Filtration and water regulation are an intimately linked processes in the kidney and regulated in units called nephrons in the kidney. Kidneys can hold millions and millions of different nephrons. Within these nephrons, the process of filtration and water regulation begins in the glomerulus, a

Balancing a Sauropod
https://doi.org/10.1016/B978-0-12-823303-0.00007-0

network of capillaries encased within Bowman's capsules. Capillary blood enters the glomerulus through small arteries called afferent arteriole and exits through equally small arteries called the efferent arteriole. The glomerular filtration unit consists of three layers: endothelial cells, a layer of acellular matrix underneath the endothelial cells, and podocytes. This structure allows water, ions, and small molecules to pass into Bowman's capsule while preventing larger molecules such as proteins and RBCs from crossing. The resulting filtrate, similar to plasma but lacking significant proteins and cells, marks the initial stage of urine formation.

After filtration, the filtrate moves into the proximal convoluted tubule, where significant reabsorption of essential substances occurs. Over half of the filtered water and ions, as well as nearly all glucose and amino acids, are reabsorbed here. This reabsorption is mediated by various transport proteins and channels, ensuring the recovery of plasma nutrients. The proximal convoluted tubule also secretes waste products into the tubular fluid. In mammals, the loop of Henle, following the proximal convoluted tubule, plays a crucial role in concentrating urine. Its descending limb is permeable to water but not to solutes, leading to water reabsorption and increased filtrate concentration. The ascending limb, in contrast, is impermeable to water but actively transports Na^+, K^+, and Cl^- out of the filtrate, decreasing its concentration.

The filtrate then enters the distal convoluted tubule, where further reabsorption and secretion fine-tune the composition of the urine. The distal convoluted tubule is sensitive to hormones such as aldosterone secreted by the adrenal gland in response to the renin angiotensin-aldosterone system (described further) or elevated plasma K^+ levels, which increases sodium reabsorption and potassium secretion. The final adjustments to the filtrate occur in the collecting duct, influenced by antidiuretic hormone released by the posterior pituitary gland in the brain in response to increased plasma osmolality or decreased blood volume. Antidiuretic hormone enhances the duct's permeability to water, allowing more water to be reabsorbed, concentrating the urine increasing blood volume and therefore, blood pressure as the volume increases the transmural pressure (per Chapter 4). This ability to reabsorb water is important for the animal's fluid balance and blood pressure, which bring us to another crucial component to the kidney.

The kidney 2: Renin-angiotensin-aldosterone system

The renin-angiotensin-aldosterone system regulates blood pressure, fluid, and electrolyte balance. The renin-angiotensin-aldosterone system is initiated in the glomerulus again, back at the afferent arterioles. As mentioned, these are small arteries, and like we learned in Chapter 4, arteries consist of endothelium and smooth muscle surrounding them to contract. In the afferent arterioles, just before the glomerulus, the smooth muscle cells become a very specialized cell type called juxtaglomerular cells. The juxtaglomerular cells have a hormone contained within them called renin, and in response to low blood pressure in these arteries, the renin is released. Renin can also be released due to decreased sodium chloride in the distal tubule or sympathetic nervous system activation in the kidney. The renin hormone catalyzes the conversion of angiotensinogen, a liver-produced protein, into angiotensin I. Angiotensin I itself is inactive, but serves as a precursor to the more potent angiotensin II.

Angiotensin I is converted into angiotensin II by the enzyme angiotensin-converting enzyme, predominantly found in the lungs and endothelium of blood vessels. Angiotensin II exerts several effects directly on the kidneys and systemically to increase blood pressure and restore fluid balance. In the kidneys, angiotensin II constricts efferent arterioles, which increases glomerular filtration pressure and enhances sodium and water reabsorption in the proximal convoluted tubule. This helps to maintain glomerular filtration rate even during low blood flow conditions, ensuring that the kidneys continue to filter blood effectively.

Angiotensin II also stimulates the adrenal cortex to release aldosterone, a hormone that increases sodium reabsorption and potassium excretion in the distal convoluted tubule and collecting ducts. This sodium reabsorption draws water back into the bloodstream, increasing blood volume and pressure. In addition, angiotensin II triggers the release of antidiuretic hormone from the posterior pituitary gland, promoting water reabsorption in the collecting ducts. Together, these actions of the renin-angiotensin-aldosterone system tightly regulate kidney function, ensuring adequate blood pressure and fluid balance, which are essential for overall homeostasis.

The kidney 3: Even more roles for this organ

Another role for the kidneys is the excretion of metabolic waste products, including urea, creatinine, and various other byproducts. Urea, produced from the metabolism of amino acids, is filtered in the glomerulus and partially reabsorbed in the tubules. Creatinine, a by-product of muscle metabolism, is filtered and minimally reabsorbed, making it a reliable indicator of kidney function.

Like the respiratory system discussed in Chapter 3, the kidneys also help to regulate acid and base balance. They do this by reabsorbing bicarbonate and excreting hydrogen ions. The proximal tubule reabsorbs the majority of bicarbonate, while the distal tubule and collecting duct secrete hydrogen ions, which combine with urinary buffers such as phosphate and ammonia to form excretable acids in the urine. This process helps neutralize and eliminate excess acids generated by metabolism and maintain blood pH within the narrow range previously discussed for normal cellular function.

The kidney 4: Giraffe adaptions = sauropod adaptions?

The kidneys have several protective mechanisms to prevent damage and maintain their function. Autoregulation, noted in Chapter 4, is a means by which higher pressure contracts an artery through the smooth muscle (this is a myogenic response). Here the autoregulation ensures a consistent glomerular filtration rate despite fluctuations up or down in blood pressure. Additionally, the kidneys possess robust antioxidant systems and repair mechanisms to mitigate damage from reactive oxygen species and other insults. These are built in protection mechanisms for an animal whose blood pressure may fluctuate from the normal range we had mentioned in Chapter 5 across most species at approximately 100–120 mmHg. However, the sauropod was fighting hydrostatic pressure, as we previously discussed, and likely had quite a large blood pressure.

For this reason, we again we return to our friend the giraffe for an example of an animal to compare with the sauropod (with our previously mentioned caveats). How do their giraffes work compared with other animals? First off, giraffes exhibit lower glomerular filtration rate and effective renal plasma flow compared with mammals of similar size, despite maintaining similar filtration fractions. Glomerular filtration rate is a measure of how much blood flows through the glomeruli over a certain amount of time. The low glomerular filtration rate is surprising given the high arterial pressure perfusing their kidneys, which typically increases filtration pressure and glomerular filtration rate. For example, the glomerular filtration rate of a 500-kg giraffe is significantly lower than predicted for its size, at approximately 0.7 mL/kg/min, compared with the expected 1.2 mL/kg/min. This is notably lower than glomerular filtration rate values observed in similarly sized animals such as horses. The low glomerular filtration rate in giraffes may suggest that they do not conform to typical mammalian scaling laws, or it may indicate a lower metabolism for the giraffe.

This anomaly is partially explained by the giraffes' unique adaptations: a thick and durable renal capsule that sustains high interstitial hydrostatic pressure in the Bowman's capsule and a valve structure at the junction of the renal vein and vena cava, maintaining high renal venous pressure. Biomechanical analysis of giraffe kidneys reveals a significantly stronger renal capsule compared with cows, capable of withstanding intrarenal pressures up to 650 mmHg!!

Furthermore, the renal resistance index, a measure of systolic and diastolic blood velocity in the kidney, in giraffes is much lower than that in humans and other animals, suggesting lower vascular resistance. This is likely part of a broader trend linking the animal's high mean arterial pressure with low resistance index. Anatomical differences in the kidney likely help with this, including how the caudal vena cava near the renal vein entry is arranged suggests unique hemodynamic properties that could mitigate the pressure. These arrangements in the giraffe kidney may serve as a starting point for understanding how the sauropod's likely high blood pressure was able to maintain kidney function, a point discussed further in the last chapter.

The kidney 5: Further adaptions for water conservation

One of the most common adaptions a kidney can have, especially in mammals, is their medullary thickness; the medulla is where the loop of Henle is stored, and so this is a direct readout of the ability of the animal to reabsorb water and concentrate their urine. The relative medullary thickness generally reflects the animals' environmental and water conservation needs. Relative medullary thickness measures the thickness of the kidney medulla relative to the overall kidney size. Animals in arid environments, such as desert-dwelling species, exhibit higher relative medullary thickness values. This adaptation is essential for maximizing water reabsorption and producing highly concentrated urine, enabling these animals to survive with minimal water intake. For instance, camels possess significantly thicker medullas compared with animals inhabiting regions with normal rainfall.

In desert-dwelling animals, this physiological adaptation is essential. For example, kangaroo rats can produce urine that is several times more concentrated than their blood plasma, allowing them to extract the maximum amount of water from their food and reduce water loss. Conversely, animals in aquatic or moist environments, such as beavers or certain amphibians, have a lower relative medullary thickness. These animals do not face the same water conservation pressures and, therefore, produce more dilute urine. Their kidneys are adapted to excrete excess water efficiently, which helps maintain electrolyte balance without the need for high urine concentration. The variation in relative medullary thickness across species highlights the diverse strategies animals have evolved to manage water and electrolyte balance, ensuring their survival in a wide range of habitats.

The differences in relative medullary thickness among species illustrate the intricate relationship between kidney structure and environmental adaptation. The ability of kidneys to concentrate or dilute urine is a direct response to the availability of water in an animal's habitat. This physiological trait is a result of evolutionary pressures that have shaped the kidneys to optimize water and electrolyte balance, ensuring the survival of species in diverse environments. Would the sauropod also have such an adaption? Although more recent evidence seems to indicate sauropods in more savannah-type environments (see Chapter 2), there are hypothesis that sauropods still stuck close to water sources/ocean. Or was the sauropod kidney evolved enough to have these types of adaptions? We look to modern-day lizards for examples.

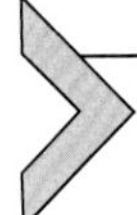

The kidney 6: Devolving the kidney; was the sauropods kidney just less complex?

The evolution of the kidney appears to begin in a similar pattern across all vertebrates' embryos, but depending on the species, the kidneys only advance to particular levels of development. The very primitive kidney systems in embryos, which survive into adulthood in only a few primitive fish species, are called pronephros. In adulthood, these types of kidneys are characterized as having one or very few rather large nephrons that filter blood at a single glomeruli.

The next stage during development of the kidney is a mesonephros type of embryonic kidney found in vertebrates that functions during early development. It consists of a series of segmented nephrons arranged longitudinally along the dorsal side of the embryo. Each nephron includes a glomerulus, Bowman's capsule, and a tubule, which drains into the mesonephric duct leading to the cloaca or bladder in some species. This arrangement supports essential filtration and excretion. In certain vertebrates, such as fish and amphibians, the mesonephros remains functional throughout their lives, adapting as their primary excretory organ. In these species, the mesonephros is efficient and supports their aquatic or semiaquatic lifestyles. However, in higher vertebrates, the mesonephros regresses as the metanephric kidney develops, taking over excretion and regulation of fluid balance. Remnants of the mesonephric tubules in these higher vertebrates sometimes differentiate into components of the reproductive system.

In reptiles, the metanephric kidney makes its debut. The reptilian kidney exhibits a greater degree of segmentation compared with that of amphibians and fish, although it still lacks a distinct cortex and medulla. Each nephron within the lobules is aligned perpendicularly to the kidney's long axis and connects to the collecting ducts. Structurally, the metanephric kidney is composed of numerous nephrons, each containing a renal corpuscle (glomerulus and Bowman's capsule) and a tubular system (proximal tubule, distal tubule, and collecting duct). In lizards and other reptiles, the metanephric kidney adapts to their environment by balancing water conservation and excretion. Due to the absence of a loop of Henle, these kidneys cannot produce highly concentrated urine. Instead, most reptiles conserve water through behavioral adaptations, such as seeking shade and burrowing to reduce water loss.

However, there are several instances where reptiles need to excrete salt to maintain fluid balance, and since they are lacking a loop of Henle, this would make fluid balance difficult. In order to excrete the excess salt reptiles may produce—either by living in or near the ocean, or the desert environment and lacking adequate shade—they have evolved salt glands that directly secrete excess salt. In many reptiles, salt glands have actually replaced where the nasal glands are normally found. Salt glands can also be found near the rectum for easy excretion as well. The salt glands are highly innervated with vasculature and cholerinergic nerves, so it's likely that the salt glands possess osmoreceptors to monitor salt, similar to baroreceptors found in the vasculature to monitor pressure. However, these glands are still not well described in terms of their molecular underpinnings.

Similar to reptiles, birds develop a metanephric kidney. However, unlike the kidneys of reptiles and other nonmammalian vertebrates, avian kidneys have distinct cortical and medullary regions, akin to mammalian kidneys. This distinction arises from the evolutionary development of Henle's loop, which creates a radiating pattern around the collecting ducts and vasa recta, forming the relatively small medullary regions in birds. In birds, larger nephrons feature a highly convoluted proximal tubule and well-developed Henle's loops, while most peripheral nephrons are short, nonconvoluted, and lack Henle's loops, arranged around large efferent veins to form the cortex.

Thus, although reptiles had similar metanephric kidneys similar to birds and mammals, there is a difference in their kidneys focused mainly on the ability to concentrate salt. Where the sauropods stood on this continuum will be discussed in the last chapter, but it is likely they would have probably fallen somewhere in between.

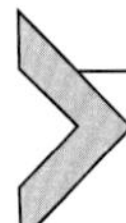

Reniculate kidneys: A possible indicator of sauropod environment?

Clearly, we don't know what type of kidney a sauropod had, but an important function of their kidney would have had to be to retain water. For that reason, the last type of kidneys we will examine are highly specialized, are called reniculate kidneys, and are primarily found in marine mammals such as whales, seals, and dolphins. However, certain terrestrial animals, such as some species of otters and the platypus—including bears!—also possess reniculate kidneys. The common thread between these animals is that they live in environments where efficient filtration and water conservation are important components to homeostasis.

Reniculate kidneys are composed of numerous small, discrete lobes, or reniculi, each functioning like a mini-kidney. Other animals such as cattle have multiple lobes of the kidney, but that doesn't alter the function. The reniculate anatomical adaptation allows these animals to efficiently filter their blood and produce urine, despite the high salt content of their marine environment. The multiple reniculi increase the surface area for filtration and reabsorption, enabling these animals to manage their water and electrolyte balance more effectively than if they had a single, large kidney.

The basic structure of reniculate kidneys is unique in that it provides a high degree of redundancy. Each reniculus has its own cortex and medulla, along with its own blood supply, which means that if one part of the kidney is damaged, the other smaller reniculate kidneys can continue to function. This redundancy ensures homeostasis if one of the reniculate kidneys is injured—with filtering sea water, this can be an issue. Additionally, the increased filtration surface area allows for more efficient removal of waste products and excess salts from the bloodstream, which is essential for saltwater-type animals that do not live in freshwater.

In addition, reniculate kidneys are especially well-suited for high metabolic demands of marine mammals especially because they need to conserve water but continually excrete salt that is absorbed from the environment in which they live. The unique architecture of these kidneys facilitates the

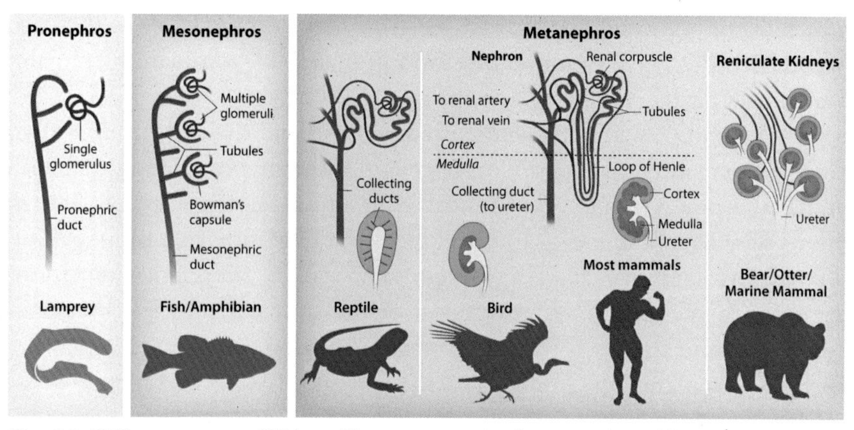

Fig. 6.1 Different types of kidney-like structures in the animal world.

production of highly concentrated urine, which helps to conserve body water. This is also likely why the bears have reniculate kidneys is that during hibernation, their urine is highly concentrated. It also may help the polar bear whose high saltwater lifestyle places it in the same category as whales and dolphins. The ability to produce concentrated urine minimizes water loss, ensuring that these animals remain hydrated even when living in a saltwater environment. The different types of kidneys discussed above are synthesized in Fig. 6.1.

Metabolism 1: Extracting the energy

We have focused extensively on the detailed regulation of the cardiovascular system in individual chapters and rounded out that discussion with integrating kidney function. In all of these cases, we have compared across organisms in an attempt to figure out pieces of the physiology puzzle the sauropod may have utilized to thrive. In this last section, we discuss a much more highly conserved mechanism that is equally important to the sauropod physiology: metabolism. Extracting energy from ingested food for metabolism is an important homeostatic mechanism because it provides the energy for muscles to walk, air sacs to inhale, heart to beat, and arteries to contract.

The way in which food from plants is broken down and utilized by animals is quite dependent on the type of animal and the environment niche in which they thrive and survive—from mammals to lizards and birds, they each use something different. Here we begin by describing four ways in which digestion occurs and try to assess what may fit more likely to the sauropod. Broadly, we'll examine digestion in mammals, ruminant mammals, lizards, and birds.

We begin by examining omnivore mammals and then compare with a range of animals that primarily eat vegetations (herbivores), along the lines of the sauropod. Ominovore mammals have a monogastric stomach, which means they have a single-chambered stomach. Digestion in a monogastric stomach involves mechanical and chemical processes. Mechanical digestion begins with the churning movements of the stomach muscles, breaking down food into smaller particles. Chemical digestion is facilitated by gastric juices, which include hydrochloric acid and digestive enzymes such as pepsin. Pepsinogen, secreted by the stomach lining, is converted into pepsin in the acidic environment. Pepsin is crucial for protein digestion, breaking down complex protein molecules into smaller peptides that can be further digested in the small intestine. Additionally, the stomach secretes mucus to protect its lining from the corrosive effects of the acid and enzymes. The rate at which the stomach empties its contents into the small intestine is controlled by the pyloric sphincter. This process ensures that the chyme (partially digested food) is released slowly and steadily, allowing for optimal digestion and absorption in the small intestine. This is simple and efficient process to digest food for energy.

Rumination is a physiological process typically associated with ruminant mammals, such as cows, sheep, and giraffes, which have a complex stomach structure with multiple chambers designed for fermenting tough plant materials with the help of microbes in the *foregut*. This process involves regurgitating partially digested food, rechewing it, and swallowing it again to enhance digestion.

We will start with the giraffe due to its long neck similarity with the sauropod. Giraffes start the digestion process by stripping leaves from trees and using their molar teeth to grind the leaves. This mechanical breakdown increases the surface area for enzymes to act on later. The chewed leaves are swallowed and pass down the esophagus into the first of a four-chamber stomach called the rumen. The rumen is a large fermentation vat where microbial digestion begins. Bacteria, protozoa, and fungi break down the cellulose and hemicellulose in the plant cell walls, producing volatile fatty acids, which are absorbed in the blood and are used as a primary energy source. Even if sauropods use different digestion, imagine how long food would take to get down the sauropod neck! This swallowing mechanism would need to ensure that food doesn't get stuck.

Periodically, the giraffe regurgitates partially digested food, known as cud, back into the mouth to chew it again. This process is known as rumination and helps to further break down the food, increasing the efficiency of microbial fermentation. After rechewing, the cud is swallowed again and returns to the rumen for further fermentation. From the rumen, the food moves into the second chamber called the reticulum, another fermentation chamber, which further aids in breaking down complex carbohydrates. The food then passes into the third chamber, called omasum, which acts as a filter. The omasum absorbs water and the fatty acids, reducing the volume of ingested food. The final chamber is called the abomasum and functions similarly to a monogastric stomach described earlier. It secretes hydrochloric acid and digestive enzymes such as pepsin, which break down proteins. This acidic environment kills the microbes carried from the rumen, allowing the giraffe to digest them and utilize their proteins. As digested food moves to the large intestine, the remaining indigestible material moves into the large intestine. Here, water and electrolytes are reabsorbed, and the waste is formed into feces. The cecum and colon also house a population of microbes that ferment any remaining undigested fiber, producing additional fatty acids that can be absorbed by the giraffe. The giraffe loses very little of the nutritional component from the low-nutrition leaves in this process. Did the sauropod have this complication of a digestive system?

The next animals we will examine to potentially understand sauropod physiology are lizards, which have a more simple digestive system. Their stomachs are not designed for fermenting plant material through microbial action like the ruminants aforementioned. Instead, lizards rely on enzymes and acids in their stomachs to break down food. Herbivorous lizards, such as iguanas, do not practice rumination. Instead, they rely on microbial fermentation in the hindgut, similar to some other nonruminant herbivores (e.g., rabbits and some mice). Microbial fermentation in the *hindgut* is a digestive process involving the breakdown of fibrous plant material by symbiotic microorganisms. This primarily occurs in the large intestine and cecum of hindgut. The process begins when fibrous plant material enters the cecum and large intestine. Here, the microorganisms secrete cellulase and other enzymes to break down the cellulose and hemicellulose into simple sugars, which are then fermented into volatile fatty acids such as acetate, propionate, and butyrate. These fatty acids are absorbed through the gut wall and serve as an energy source for the host animal.

Similar to ruminants, the microbial fermentation produces gases such as methane and carbon dioxide as by-products. The fermentation also results in the synthesis of microbial protein and vitamins, particularly B-vitamins, which can be partially absorbed by the host. However, since much of the microbial protein is not utilized by the host in hindgut fermenters, it is often excreted in the feces. This process is less efficient compared with foregut fermentation seen in ruminants, but it allows for a faster passage of food through the digestive system, which can be advantageous in environments where food is abundant but of low nutritional quality. This type of environment is much like what the sauropod likely encountered.

The bird stomach is the last mechanism of digestion we will examine across animals. Birds do not use rumination. Generally, the bird stomach consists of two main parts: the proventriculus and the gizzard (ventriculus). Starting at the proventriculus, also known as the glandular stomach, this is the first part of the bird's stomach. It functions similarly to the mammalian stomach, secreting digestive enzymes and hydrochloric acid. Also similar to a mammalian stomach, these secretions help to break down food chemically before it moves to the next part of the stomach. Of particular interest is how the proventriculus is adapted for the rapid secretion of digestive juices, allowing for a quick type of "initial" digestion. This part of the stomach is especially well-developed in carnivorous birds.

Next is the gizzard, or ventriculus, which is a highly muscular organ that acts as a mechanical grinder. You may have seen chickens do this, but it is common across all birds: they swallow small stones or grit, which are stored in the gizzard to aid in grinding food. This process is necessary because birds do not have teeth to chew their food. The gizzard's strong muscular walls contract forcefully, crushing and grinding the food, mixing it with the digestive juices from the proventriculus. This mechanical digestion is particularly important for birds that consume hard seeds or tough plant material. The structure and functionality of the gizzard vary among bird species, reflecting their diet. For example, birds such as pigeons and chickens have a more robust and muscular gizzard compared with carnivorous birds, which rely more on chemical digestion from the proventriculus.

Certainly, the sauropods of the Jurassic were herbivores (although this isn't clear for some of the early sauropods in the Triassic), but what type of digestion did they have? What we do know is that several sauropod gastroliths have been identified. These gastroliths, ranging in size from pebbles to boulders, were likely integral components of sauropod digestive physiology. The process of gastrolith ingestion, known as gastrolithiasis, involved the deliberate swallowing of stones by sauropods like modern-day birds described earlier, but also by several herbivore lizards, alligators, and crocodiles.

Metabolism 2: Everything come out ok?

Once the food is broken down in the stomach, the intestines absorb the nutrients, but what can this tell us about the sauropod physiology? In general, longer intestines allow more time to digest and ferment bacteria from plants. For example, compared with the body size of almost any other mammal, cats have the shortest digestive tract. Since carnivores consume highly digestible raw meat, there is no need for a long digestive tract. Marine mammals tend to have especially long intestines compared with their land-dwelling counterparts, a feature that is noticeable in species such as seals, sea lions, and whales. For example, the manatee, with an intestinal length of approximately 147 ft., yes over 100 ft. long, has this adaption to allow for digesting its plant-based diet effectively. The blue whale also has impressively long intestines at approximately 98 ft. Both the manatee and blue whale are hind-gut fermenters. As one might expect, the giraffe has an exceptionally long intestine as well measuring in at a whopping 260 ft. in length even as a foregut fermenter!

At the end of this digestive tract is the feces. Fossilized feces from dinosaurs are called coprolites. These ancient feces provide valuable insights into the diet and digestive processes of sauropods. For instance, the presence of well-digested plant material might suggest a specialized herbivorous diet, while fragments of bone in carnivorous dinosaur coprolites indicate bone-crunching feeding behaviors. This dovetails into interesting work by Prasad et al. in 2005 that demonstrated in titanosaur coprolites the presence of conifers and cycads as one would expect from a sauropod and as discussed in Chapter 2. However, they also found the presence of ancient grasses in these coprolites with implications for earlier evolution of grasses, but also tantalizing implications in terms of what sauropods consumed. As we had also postulated in Chapter 2, perhaps it was possible that sauropods were also enjoying some early angiosperms with their grass. But it does beg the question about the nutritional value of each of these plants.

There is a significant difference between eating grasses and eating conifers in terms of metabolism. These differences stem from the distinct compositions and structures of these plant types, which influence their digestibility and nutritional content. Grasses are primarily composed of cellulose and lignin and require specialized digestive processes to break down these fibrous materials. Herbivores that uniquely feed on grasses, such as ruminants described earlier, have adaptations to ferment and break down cellulose into usable nutrients. Conifer needles and leaves contain not only cellulose but large amounts of secondary compounds such as resins and tannins, which can be toxic or difficult to digest. These compounds can inhibit digestive enzymes and reduce the overall digestibility of the plant material. Animals that feed on conifers often have detoxification mechanisms to deal with these chemicals.

Typically, grasses have lower protein and higher fiber content compared with conifers. However, they provide a steady source of energy through the fermentation of cellulose into fatty acids described earlier. Grasses also tend to have higher levels of essential minerals such as potassium, phosphorus, and calcium. Conversely, coniferous plants generally have higher protein content due to their evergreen needles, which contain more nitrogen. However, the presence of defensive compounds in conifers can limit the availability of these nutrients.

Metabolism 3: The sauropod energy

The next step as the food is broken down and absorbed across the intestine is called catabolism. Catabolism is a metabolic process in which the food intake is broken down into simpler ones, releasing energy. This process breaks down large molecules such as carbohydrates, fats, and proteins into smaller molecules; proteins primarily into amino acids, carbohydrates into glucose, and fats into glycerol and fatty acids. From these molecules, all animals derive energy for their cells to function and story energy for later use (e.g., in fat). Every single cell almost exclusively uses ATP to drive essential cellular processes such as creating a membrane potential, releasing hormones, or making new proteins. ATP is a simple molecule with three phosphates attached; when each phosphate is removed, or a phosphate attached, it creates the needed energy for each of these physiological processes to occur.

Let's start with glucose as an energy source as it is the main mechanism to quickly and efficiently derive ATP. Glucose is broken down in a process termed glycolysis that happens within the cytoplasm of cells. After glycolysis, one molecule of glucose has produced four ATP molecules and one molecule of a metabolic product termed pyruvate. It did take two ATP molecules for this entire reaction to occur, so the net amount of ATP was only two molecules, but that is already a good start. But the real power of ATP production comes next.

The pyruvate molecule produced from the molecule of glycolysis is transported into the mitochondria, a key organelle in all cells (especially the heart), and a small amount of the "waste" product we talked about in Chapter 3 is produced: CO_2. Once inside the cell, pyruvate is broken is enzymatically altered to produce acetyl-CoA. It's this molecule that sets off a reaction in mitochondria that generates the next large quantity of ATP. The acetyl-CoA enters the tricarboxylic acid (TCA) cycle, also known as the citric acid cycle or Krebs cycle, and is oxidized to produce even more CO_2, NADH, $FADH_2$, and ATP. The other two main molecules produced during that initial catabolic step, amino acids, and fatty acids, also feed into the TCA cycle; fatty acids can directly produce acetyl-CoA by beta oxidation, and amino acids are key requirements for several of TCA enzymes.

The ATP can be readily used, and the NADH and $FADH_2$ generated in this process can then be used in the electron transport chain on the mitochondria to produce even more ATP through oxidative phosphorylation. For these reasons, mitochondria are labeled as "powerhouse" of the cell due to its localized role in producing ATP. This is also why the amount of CO_2 being produced by cells is an indirect measurement of how metabolically active cells are, as it means pyruvate is moving to the mitochondria and the TCA cycle is active. It is also important to note that this process of metabolism produces heat from the chemical reactions, especially as millions of cells become more metabolically active, there is significant heat generation the animal must dissipate.

If the glucose, fatty acids, or amino acids are not used for catabolism, then their potential use for energy generation is implemented, a process termed anabolism. Anabolism is a metabolic process that involves the synthesis of complex molecules from simpler ones. This process is crucial for growth, repair, and maintenance of tissues in the body. During anabolism, small molecules such as amino acids, monosaccharides, and fatty acids are built up into larger and more complex molecules such as proteins, polysaccharides, and lipids. For example, fatty acids are converted into lipids for storage as fat, and amino acids are converted into muscle. The fat is a particularly rich energy source for storage as it can be broken down as needed into fatty acids again and then enter catabolism through conversion to acetyl-CoA.

Metabolism 4: Storing the sauropod energy

The storage of fat (adipose) in animals by anabolic processes noted earlier is crucial for survival, especially in difficult environments where food may be scarce. The distribution of the fat in the animal is also important, as its localization would ideally be centered closest to where the energy would be needed the most or quickest. Let's discuss how fat is distributed across a few different species and see if we can understand how the sauropod may have done this.

Lizards store adipose tissue primarily in three areas: their tails, the abdomen, and sometimes under the skin. The tail especially serves as the most significant energy reserve, especially vital for periods of food scarcity, hibernation, or high energy demands such as reproduction. In the coelomic cavity, fat is stored near the reproductive organs and intestines, providing crucial metabolic support during fasting or reproductive activities. The distribution and amount of adipose tissue can vary widely among different species and are influenced by factors such as habitat, diet, and seasonal changes. For example, desert-dwelling lizards might rely more heavily on tail fat stores due to the extreme variability in food availability. These fat reserves ensure they have the necessary energy to endure times of scarcity and support vital physiological functions.

Birds store fat in various parts of their bodies, primarily as adipose located just beneath the skin. These fat reserves are particularly concentrated in certain areas such as the abdomen and the base of the wings. Additionally, birds accumulate fat within their abdominal cavity around internal organs, including the intestines and liver. Fat storage is crucial for birds, especially during periods of migration, breeding, and molting, when energy demands are significantly higher.

Giraffes store fat in several parts of their bodies, mainly in subcutaneous deposits located under the skin, much like other mammals. These fat deposits are particularly found around the base of the tail, shoulders, and the areas surrounding the joints, such as the hips and knees. Additionally, giraffes can accumulate fat around their internal organs, particularly within the abdominal cavity.

There are, however, other reasons for where animals, and our sauropods, could store fat. For example, camels have humps of fat on their back and nowhere else. The reason for this isn't because muscles on the back are in particular need of the stored energy any quicker than any other muscles, but for thermoregulatory reasons. It cools down the camel having fat in one place on their body, and the ongoing metabolism in the gut that produces the vast amount of heart is easily dispersed. Remember that metabolism is a huge heat source, so dissipating it quickly and easily is an important thermoregulatory mechanism. Based on the information above, we can start

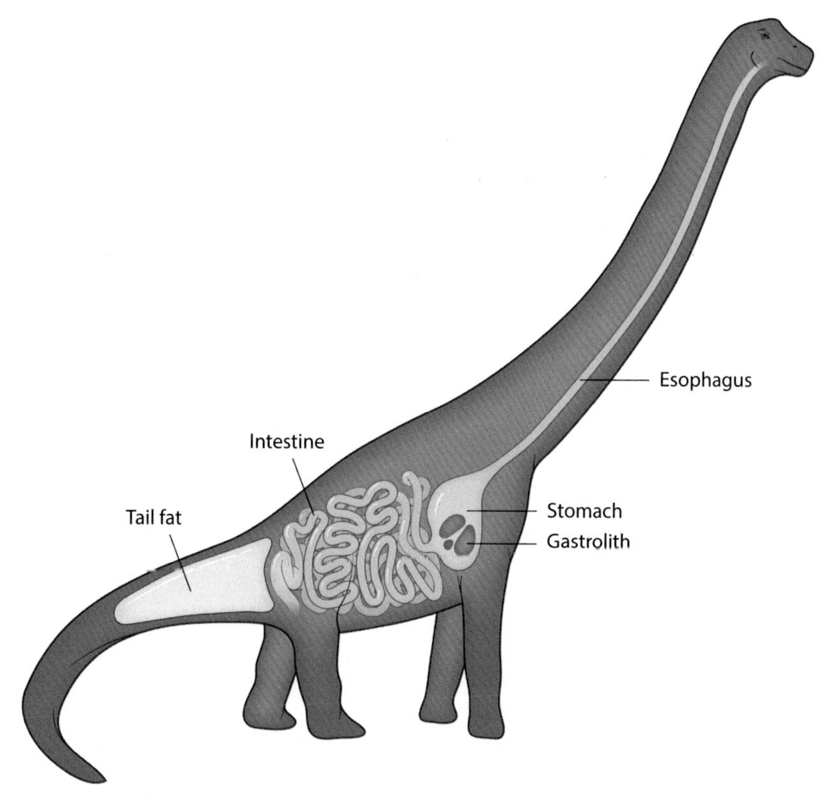

Fig. 6.2 Simplified diagram of the possible organization of the digestive system of the sauropod.

to picture what the sauropod digestive system may have looked like in simplified terms (Fig. 6.2).

Overall, the kidney and metabolism are important links to respiration, vasculature, and heart function that together with trace and body fossils from sauropods and the plant environment in which the sauropods lived, could together shed light on this amazing creature. For these reasons, lets finish the book by attempting a big picture synthesis of what the sauropod may thrived and survived in the Mesozoic.

References and further reading

Boron WF, Boulpaep EL: *Medical physiology*, ed 2, New York, 2009, Elsevier.

Curry Rogers KA, Wilson JA, editors: *The sauropods*, Berkley and Los Angeles, California, 2005, University of California Press.

Damkjær M, Wang T, Brøndum E, et al.: The giraffe kidney tolerate high arterial blood pressure by high renal interstitial pressure and low glomerular filtration rate, *Acta Physiol* 214:497–510, 2015.

Hallett M, Wedel MJ: *The sauropod dinosaurs*, Baltimore, Maryland, 2016, John Hopkins University Press.
Hill RW, Wyse GA: *Animal physiology*, ed 2, New York, New York, 1989, Harper Collins Publishers.
Holz PH: Anatomy and physiology of the reptile renal system, *Vet Clin North Am Exot Anim Pract* 23:103–114, 2020.
Homgren S, Olsson C: Autonomic control of glands and secretion: comparative view, *Auton Neurosci* 165:102–112, 2011.
Klein N, Remes K, Gee CT, Sander PM, editors: *Biology of the sauropod dinosaurs*, Indianapolis, Indiana, 2011, Indiana University Press.
Koeppen BM, Stanton BA: *Berne and Levy physiology*, ed 6, Philadelphia, Pennsylvania, 2008, Mosby Elsevier.
Mohrman DE, Heller LJ: *Cardiovascular physiology*, ed 5, New York, New York, 2003, McGraw-Hill.
Prasad V, Stromber CE, Alimohammadian H, Sahni A: Dinosaur coprolites and the early evolution of grasses and grazers, *Science* 310:1177–1180, 2005.
Romagnani P, Lasagni L, Remuzzi G: Renal progenitors: an evolutionary conserved strategy for kidney regeneration, *Nature Rev Neph* 9:137–146, 2013.
Seifert G, Sealander A, Marzen S, Levin M: From reinforcement learning to agency: frameworks for understanding basal cognition, *Biosystems* 235, 2024 105107.
Seymour RS: Cardiovascular physiology of dinosaurs, *Physiology (Bethesda)* 31:430–441, 2016.

CHAPTER SEVEN

Balancing the sauropod

As noted in the chapters at the beginning, homeostasis involves fluid balance and pH balance; these are the core basis of physiology that all living things, from plants to all animals, must abide by. Balancing these core basics are necessary for an animal to live day to day. So what was it about the sauropod physiology that kept it in balance? Why did it thrive and how did it thrive? Physiology can help us understand that.

If we were to consider the sauropod as a machine, the sauropod had two major design issues, its neck and its size. The sauropod was the largest land animal to ever live on earth and the animal with the longest distance from the heart to the brain. These weren't simply "flukes"; evolution doesn't allow flukes to persist and radiate for 150 million years. The sauropod size and neck length were essential characteristics that make it a unique clad of species. Before we speculate further on these two species hallmarks, let's take a moment to review the core physiology concepts covered in this book.

I strongly focused our physiology discussion broadly on the cardiovascular system as these factors would likely have to have the most adaptions for the sauropod to be in balance—from breathing the oxygen-poor air to moving blood up the neck and digesting enough food for energy. We will review the key take aways from these detailed chapters and specifically focus on what they may mean to sauropod physiology. For each of the following sections, I present what I hypothesize the sauropod physiology actually was. Between the review of the key points and the detailed physiological discussion in the previous chapters, I hope there is sufficient rationale for my hypothesis. But like all of science, hypotheses are made to be broken, especially as new evidence is discovered, and new ideas are rounded out.

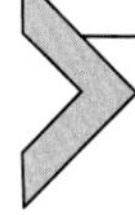

Hypothesis: Sauropods had small, nucleated red blood cells

As previously mentioned, I made an argument that possibly one of the most important cells of a sauropod was its red blood cells (RBCs). These cells are integral to understanding an animal's physiology and environmental adaptations due to their primary function in oxygen transport to tissues

Balancing a Sauropod
https://doi.org/10.1016/B978-0-12-823303-0.00011-2

for support of nearly all functions necessary for life. Thus, the RBCs, how they're transported, and how they are oxygenated are vital to understanding sauropod physiology.

Red blood cells contain hemoglobin, a protein that binds O_2 in the lungs and releases it to tissues throughout the body, a process crucial for cellular respiration and energy production. The hemoglobin protein takes up the O_2 with an iron molecule inserted into the heme part of the hemoglobin protein. It makes sense then that the maintenance of the sauropod cellular iron would have been essential to ensuring that hemoglobin was always functionally available. In fact, iron-rich sedimentary rocks (>15% iron) occurred in large amounts in three different time periods of the earth, Precambrian, Paleozoic, and the middle Mesozoic, the most abundant time of the sauropod. A massive increase in soil iron would have increased the iron levels in plants and vegetation too and may have ensured proper hemoglobin function for sauropod RBCs. This would have been especially important in the low-O_2 environment of the Mesozoic.

This O_2 delivery system also supports metabolic activities necessary for survival, growth, and reproduction. In environments with varying O_2 availability, such as the Mesozoic, the efficiency of O_2 transport by RBCs becomes even more critical. In that regard, we also discussed how adaptations in RBC morphology, hemoglobin affinity, and the overall count can provide insights into how animals have evolved to optimize O_2 uptake and delivery under specific environmental conditions. For instance, animals living in high-altitude environments often exhibit increased RBCs counts or modified hemoglobin with a higher affinity for oxygen to compensate for lower atmospheric oxygen levels. We had two key examples of RBC size differences: the Amphiumas and the mouse deer. The Amphiumas are amphibians with the largest size of RBCs that live in exceptionally low-O_2 environments with a slow metabolism. Presumably the exceptionally large size RBC provides the absolute maximum size to hold oxygen with all of its space for hemoglobin; however, the massive size made the blood viscous, slowing down delivery of oxygen. Conversely, the mouse deer with its extreme metabolic activity has the smallest size RBCs and no nucleus to provide the most efficient rapid transport of oxygen to the tissue. To move these tiny RBCs throughout the body, mouse deer have exceptionally high heart rates, something that would have been difficult for our sauropod.

We also learned about the presence or absence of nuclei in RBCs. A red blood cell with a nucleus could provide several evolutionary advantages. Unlike typical mammalian RBCs, which lack nuclei to maximize space for hemoglobin and thus oxygen transport, nucleated RBCs found in other vertebrates, such as birds and reptiles, retain the ability to divide and repair themselves. The nucleated RBC in sauropods could also be advantageous in environments where oxygen levels fluctuate significantly, as nucleated cells can rapidly respond to changing conditions by producing more red blood cells through mitosis. In birds and especially in sauropods, this would be crucial since there was no bone marrow space to make new RBCs through hematopoiesis as described in Chapter 1 (this doesn't mean their spleen could have been active, just that they likely lacked the ability to overproduce RBCs as needed). Additionally, the presence of a nucleus allows for the synthesis of proteins and other cellular components, which can enhance the cell's resilience to damage and increase its lifespan. This adaptability and durability could be particularly beneficial for species, such as the sauropod, living in harsh or variable environments, offering a greater degree of physiological flexibility and resilience.

For these reasons, we came down on the hypothesis that the sauropod had (in order of likelihood): 1. nucleated RBC, 2. RBC in sizes that were comparable with reptiles and birds in current day, and 3. a slightly higher hematocrit than other comparable dinosaurs. These adaptions would have allowed sauropod flexibility to respond to oxygen fluctuations with more RBCs overall (meaning greater hemoglobin availability), mitosis of RBCs due to the lack of bone marrow in their bones, cellular repair due to the distance RBCs would have had to travel over the sauropod network, and RBC size allowing for efficient movement through the vasculature. Next, let's layer on how these RBCs would have worked in their lungs.

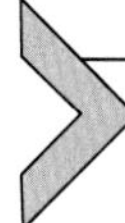

Hypothesis: Sauropods had avian lungs extending up the neck

Understanding how an animal breathes is fundamental to comprehending its physiology and ecological niche because respiration is directly linked to metabolic rate, energy expenditure, and survival strategies. The respiratory system facilitates the intake of O_2 and the expulsion of CO_2, which is crucial for cellular respiration and energy production. As we learned, different environments present unique respiratory challenges, such as variations in O_2 availability and physical demands. For instance, aquatic animals have developed gills or specialized lung structures to efficiently extract oxygen from water, whereas terrestrial animals have lungs adapted for air breathing.

Broadly, the terrestrial lungs can be divided between a more mammalian style of breathing air in (inhalation), letting the inhaled oxygen diffuse to the RBCs in the capillaries of the lung, and letting the waste CO_2 diffuse out. This same inhaled air is then exhaled with the CO_2. Although this is a rapid mechanism of O_2 intake, it is not as efficient as avian lungs because the single inhale and single exhale leave dead air space in the lungs sacs (alveoli). This pushing and pulling of the lungs to push the air in and out requires skeletal muscle movement of the diaphragm. Conversely, there is no wasted air in avian lungs. As detailed in Chapter 3, avian lungs have four cycles for one breath; these are composed of two inhalations and two exhalations. As oxygen-rich air moves down the trachea, it is stored in the posterior air sac (first inhalation), from there, actions of the rib cage can help to contract the air sacs and push the oxygen-rich air up over thousands of parabronchi. These parabronchi have capillaries laying perpendicular over them carrying oxygen-depleted RBCs to maximally extract all the O_2 from the air (first exhalation). The oxygen-depleted air, now filled with excessive CO_2, is pushed to the anterior air sacs and likely even the pneumatic bones (second inhalation). The contraction of the anterior air sacs exhales the air out the mouth (second exhalation). This mechanism maximally extractions CO_2 from the environment.

Another reason a mammalian-type respiratory system is an unlikely mode of oxygen extraction for sauropods is understood from examining giraffes. Giraffes have mammalian-type lungs that are eight times larger than other mammal's lungs, comprising a larger portion of their abdominal cavity. The relatively small ratio of the sauropod abdomino-thoracic cavity compared with body mass limits the space available for lung expansion and/or for accommodating the digestive tract. For an animal of its size, this simply won't work to digest both the quantity and low quality of food needed (see further).

In addition, logical anatomical and physiological reason for use of air sacs is that they are not confined to the abdominal cavity, and the placement of the air sacs, especially the posterior air sacs, along the neck of the sauropod could lessen the overall weight of the neck. A neck of this length that is engorged with skeletal muscle would not only be quite heavy and thus difficult to lift, it would require significant metabolic energy. Using air sacs along the neck length would lessen the load on the neck range of motion and provide sufficient body space for digestion.

For the aforementioned reasons, I hypothesize that the sauropod had (in order of likelihood): 1. avian-style lungs, and 2. their avian-style lungs ran along their necks.

Both adaptions would have provided maximum O_2 extraction for the sauropod and lowered metabolic load in terms of the weight of the neck and the lack of a need of a diaphragm. The sauropods were not stationary animals, with evidence of their sometimes-rapid movement. Therefore, even a little bit of rapid movement for such a large animal would have required maximum oxygen extraction, and the avian lungs provide just such a mechanism.

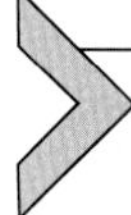

Hypothesis: Sauropods had low-resistance, high-pressure arteries like a giraffe

Understanding how an animal's vasculature works is crucial to understanding their overall physiology for several reasons. First, the vascular system is responsible for the distribution of oxygen, nutrients, hormones, and the removal of waste products throughout the body, playing a central role in maintaining homeostasis. The efficiency and capacity of an animal's vasculature can indicate its metabolic rate and energy demands, which are closely tied to its activity levels, growth rates, and reproductive strategies. For instance, animals with highly efficient cardiovascular systems, such as birds and mammals, can sustain high levels of activity and have higher metabolic rates compared with those with less efficient systems. This adaptation is often reflected in their ability to exploit a wider range of ecological niches and undertake activities such as long-distance migration, predation, or rapid escape from predators.

One of the key points about the function of the vasculature we discussed in Chapter 4 was the relationship between pressure, resistance, and flow. We made the strong assumption that for blood to be pushed up the neck and perfuse the brain, the blood pressure would have to be quite high in the arteries, especially with the hydrostatic pressure exerted on the arteries by gravity not only by the height and size of the sauropod, but especially the long neck as the blood attempted to reach the brain at the end. Because of this exceptionally high pressure, something else about the relationship between pressure, flow, and resistance would have to be altered; like the giraffe, we hypothesized the relationship that would be changed would be a low resistance system. This would ensure that blood flow was adequate to reach the tip of the tail to the legs to the end of the neck with high pressure and low resistance. We also discussed the effect of the high pressure on the blood vessels exerting a high transmural pressure, which in many situations would induce leak out of the arteries. Giraffes compensate for this with extra thick arteries, something akin to a compression sock. Part of the reason for this was a lack of skeletal muscle in their legs. However, the huge legs of a sauropod, while also serving an important purpose in holding their weight, also were likely quite muscular, helping to drive the excess fluid back to the heart.

We also discussed two ideas on how the vasculature in the neck may have worked to ensure adequate supply of blood to the brain, including the siphon hypothesis and the valve hypothesis. Although the siphon theory does account for the laws of physics and would solve the problem of getting blood up the neck from a purely logistical point of view, it is hard to view this in a physiological sense of something the animal processed. The major point is that the vasculature would have to have been developed almost completely unique to the siphon theory of blood movement of neck with severe bouts of hypoxia and potentially collapsed vessels. Although this doesn't mean it didn't happen, it would have been unique. Instead, we look at the valves in the veins as the blood returns to the heart, and we hypothesize a combination of increased number of valves and a rete mirabile at the base of the brain that may have lessened the pressure as the blood approached the brain and allowed for a counter current heat release mechanism.

Animals have also adapted unique mechanisms in the vasculature to adapt to low O_2 or short bouts of low O_2. This is especially seen in diving mammals, such as seals and whales, which have developed unique adaptations such as increased myoglobin in muscles to store O_2. This is also seen in large arteries of other mammals, although not necessarily examined in other species. The expression of myoglobin in the blood vessel walls would have been an important and easy mechanism to preserve O_2 in tissue that is distant from the oxygenated blood.

For the aforementioned reasons, I hypothesize that the sauropod had (in order of likelihood): 1. low resistance throughout the vasculature to accommodate the high blood pressure necessary to push blood up the neck, 2. significant skeletal muscle contraction in their legs to prevent edema, 3. myoglobin throughout the vasculature to capture oxygen, and 4. a rete mirabile would be present to dampen pressure at the base of the brain. As we shall see next when we discuss the heart, adaptions in the vasculature likely mirrored adaptions seen in the heart very closely.

Hypothesis: The sauropod had a smooth four-chamber heart that was hypertrophied

Understanding how an animal's heart functions is fundamental to comprehending its overall physiology and how it thrives within its ecological niche. The heart is the central organ responsible for pumping blood throughout the body, ensuring that oxygen, nutrients, and hormones are delivered to tissues while waste products are removed. The structure and efficiency of the heart can indicate an animal's metabolic demands and energy expenditure. For example, animals with high metabolic rates, such as birds and mammals, typically have more complex and efficient heart structures capable of sustaining high levels of activity. This allows them to engage in energetically demanding behaviors such as flight, long-distance migration, or rapid movement, which are critical for their survival and reproductive success. Conversely, animals with lower metabolic rates, such as reptiles and amphibians, often have simpler heart structures that reflect their more sedentary lifestyles and lower energy requirements.

The heart's adaptations provide insights into how animals have evolved to survive in specific environments. For instance, birds have hearts that can efficiently oxygenate blood despite low atmospheric oxygen levels—an adaption combining efficient oxygen extraction (described earlier), a four-chamber heart to completely separate oxygenated and deoxygenated blood, and "smooth" heart chambers to allow more laminar and less disturbed blood flow. We suppose this adaption would also benefit the sauropod to not only accommodate the low O_2, but keep oxygenated blood away from deoxygenated blood, but also accommodate the increased blood pressure, as it would also keep resistance low as it moves through the heart. Moreover, similar to giraffe and humans with heart failure, we also hypothesize that sauropods had hypertrophied left ventricles, again not only to accommodate the huge blood pressure but to push the blood up the neck. This adaption in a slightly larger heart would likely be scalable to accommodate longer distance from the heart to the brain (see further).

Another idea we posed to move blood up the neck were multiple hearts. It would be difficult for a sauropod to have extra hearts as this would be a huge metabolic investment by the sauropod to perfuse a brain with little cognitive function. Besides, the mechanisms derived by the giraffe are likely to be scalable to the sauropod as well.

For the aforementioned reasons, I hypothesize that the sauropod had (in order of likelihood): 1. a four-chamber heart, 2. a heart size in relation to its body size, 3. a heart that was hypertrophied to accommodate the increased pressure, and 4. a smooth atria and ventricles to easily move blood through the heart. The combination of these physiological adaptions would not have been a difficult mechanism to have physiological adaptions for and not required as much of a metabolic input in size comparison to that of a giraffe. However, the neck issue continues to be an issue directly related to the vasculature and heart, and we will touch on this as follows.

Hypothesis: Sauropods had a metanephros kidney akin to modern-day reptiles

The kidneys are essential for maintaining homeostasis by regulating fluid balance, electrolyte levels, and waste excretion. They filter the blood to remove metabolic byproducts, which is critical for detoxifying the body. The efficiency and adaptability of the kidneys can reveal an animal's dietary habits, water conservation strategies, and metabolic rate. For instance, desert-dwelling animals, such as camels, have highly efficient kidneys that can produce highly concentrated urine to conserve water, allowing them to thrive in arid environments with scarce water resources. In contrast, animals living in freshwater habitats, such as amphibians, have kidneys that can handle large volumes of dilute urine, reflecting their need to excrete excess water and maintain osmotic balance. Additionally, the kidneys' role in hormone production and regulation provides insights into an animal's physiological responses to environmental stressors. The kidneys produce hormones such as erythropoietin, which stimulates red blood cell production in response to hypoxia, and renin, which plays a key role in regulating blood pressure. These hormonal responses are critical for animals adapting to varying oxygen levels, such as those living at high altitudes or engaging in deep diving activities, or, as in the case with sauropods, living in low-O_2 environments.

We discussed the possibility of a salt gland that would help the kidney regulate osmotic balance. This is found in some reptiles and birds, especially those living in high salt environments, or where a more primitive kidney is found. However, more recent paleontological evidence indicates the apparent absence of this feature in more basal dinosaurs, which may indicate that it is only after dinosaur "miniaturization," i.e., modern-day birds close to the origin of flight, that excretory mechanisms were favored over exclusively renal mechanisms of salt regulation. These salt glands may have also been able to leave a bony trace due to the high salt concentration, but as of yet, there isn't any indication of salt glands in sauropods, either around the nasal cavities, eye orbitals, or around the tail.

This would suggest that the kidneys of a sauropod were likely sufficient to cope with the water balances and salt balance needed by the animal and a metanephros kidney akin to modern-day reptiles with multiple glomeruli, but lacking a medulla and cortex. However, if there were ever a soft tissue discovery of a sauropod, which is not entirely likely (but who knows?),

I might suggest that finding remnants of a kidney could be quite telling about its overall physiology. The reason for this is the debate, sometimes ongoing, sometimes apparently settled, about whether sauropods were water dwellers. We had previously mentioned in Chapter 2 that the lack of a ballast likely indicted a strong disposition against sauropods living in water. This has been supported by multiple lines of evidence, especially at the biomechanical level. Regardless of this, a more water-dwelling sauropod may have had quite different kidneys. Perhaps not true reticulate kidneys seen in bear, otters, and seals, but certainly something different than a normal reptilian kidney to handle the excessive water load for dilute urine. For this reason, the discovery of a sauropod kidney could be one of the most essential findings about sauropod physiology and ecology.

For now, *I hypothesize that the sauropod had a metanephros kidney, similar to that seen in modern-day reptiles and lacked salt glands—unless soft tissue fossils of sauropod kidneys are discovered.*

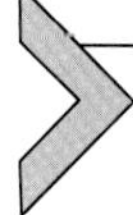

Hypothesis: The sauropod was a hind-gut fermenter with an extensive digestive track

The digestive system breaks down food into essential nutrients, which are then absorbed and utilized for energy, growth, and repair. The efficiency and specialization of an animal's digestive system reveal much about its diet and feeding strategies. For instance, herbivores such as ruminants have complex stomachs with multiple chambers to ferment and break down tough plant fibers, allowing them to extract maximum nutrients from a fibrous diet. Carnivores, on the other hand, have shorter and simpler digestive tracts optimized for processing protein-rich meat. The differences in digestive anatomy and function are critical for understanding how animals are adapted to their specific food sources and environments. In terms of sauropods, the presence of stomach gastroliths—rounded stones used to crush leaves to increase surface area for digestion—strongly suggest a digestive system similar to that seen in other reptiles and birds. The massive amount of food to maintain a metabolism to keep the heart beating for movement of blood up the neck, and the low quality of digestible plant matter, would suggest the need for an elongated digestive tract so minimal food stuffs could be fully utilized. This would also necessitate a need for hind-gut fermentation, again, something common in birds and reptiles. It is not likely they were fore-gut fermenters akin to giraffes and cows most likely due to the teeth on the sauropod fit for chewing cud and the more logistical means of moving cud up the esophagus of such an elongated neck (although giraffes do move cud up their neck).

Perhaps it was also possible that sauropods were more open to grazing across ecological niches that assumed. Evidence from fossilized titanosaur sauropod feces seems to indicate the presence of grasses, in addition to conifers, fungi, and algae. It's possible that the addition or switch to grass diets allowed titanosaur to become bigger due to the higher nutritional component. This in addition to massive gastroliths and grit form soil in their feces make it quite clear that sauropods did not live their whole life with their head stuck in one ecological niche going from tree to tree. Instead, I propose that sauropods were more grazers than originally thought, with a full range of motion of their neck. It is highly likely that their neck length gave them preferential spatial location in specific areas of trees where they may have consumed the most of food, but it didn't stop them from sampling the other delicacies of the Mesozoic.

Moreover, metabolism, which encompasses all chemical reactions in an organism, including those that convert food into energy, is closely tied to an animal's ecological role. Metabolic rates vary significantly among species and are influenced by factors such as body size, activity level, and environmental temperature. Endotherms (warm-blooded animals) such as birds and mammals typically have higher metabolic rates compared with ectotherms (cold-blooded animals) such as reptiles and amphibians. This affects their energy requirements, behavior, and habitat preferences. High metabolic rates enable endotherms to sustain prolonged activity and inhabit a wide range of environments, from polar regions to tropical zones. Conversely, ectotherms often rely on external heat sources to regulate their body temperature and may be limited to warmer climates. As mentioned in Chapter 2, there is emerging evidence that sauropods have been found away from the poles where it was cooler and closer to the equator. It is much more likely that sauropods were somewhere on the continuum, with enough conservation of heat to be active and congregate in herds, but not enough to be able to rapidly run away from predators. In addition, for the sauropod, we might predict the tail to be the most obvious area of fat accumulation due to its girth and length in most sauropod species. This is similar to modern-day lizard species (as well as crocodiles and alligators). It would also provide a mechanism to keep the body more tightly regulated in terms of heat distribution.

For the aforementioned reasons, I hypothesize that the sauropod had (in order of likelihood): 1. an extensive digestive tract that likely took up much of the abdominal body cavity, 2. stored fat in its tail, 3. hind-gut fermentation to metabolize their food, and 4. a varied diet, much more than previously thought. These four items meant they likely derived nutrition and calories from multiple food sources to keep their massive body frame moving and heart beating, although the different neck length of sauropods may have provided them dedicated food areas. The presence of sauropod-specific gastroliths further indicates hind-gut fermentation, similar to that seen in reptiles and birds. This is an area where more paleoecology evidence will be helpful in deciphering more of the sauropod physiology.

What does a sauropod think of this?

One aspect that we did not touch on is the brain and central nervous system of the sauropod. This is not for any other reason than the extreme difficulty we have in placing this into context without fossilized records, which have given us a broad overview of the likely basics of the sauropod central nervous system. As one might suspect, despite their massive size, sauropods had relatively small brains compared with their body size, which is a fairly common trait among nonavian dinosaurs. This means sauropods had what's called a low encephalization quotient. A low encephalization quotient is interpreted as a sign of limited behavioral complexity, and thus, sauropods were probably quite instinctual by nature.

The braincase of sauropods has been shown to be a thick bony structure that encased and protected the brain. It was composed of several bones, including the basioccipital, exoccipitals, supraoccipital, parietals, frontals, and the laterosphenoids. These bones together formed a rigid and well-protected brain cavity. The presence of large air-filled sinuses in the skull reduced head weight, an important adaptation for their size and something we will circle back to, especially with implications for lifting a head up, rather than keeping a head straight.

The inner ear structures, including the semicircular canals, are also preserved within the braincase. These structures are essential for balance and spatial orientation. Studies of sauropod inner ears indicate they had a good sense of balance, which was likely necessary for managing their enormous bodies. Importantly for our ongoing discussion, the orientation of the semicircular canals also suggested that sauropods held their heads in a horizontal position most of the time and thus were not very often looking down or looking up. This is yet another important data point if a sauropod is bending its neck to eat food in water, whereas eating food in front of you, for example, in a tree.

The braincase also contained many defined openings (foramina) for the cranial nerves. These nerves are considered crucial for sensory input and motor control. Notably, the olfactory bulbs (associated with the sense of smell) were relatively small, suggesting that sauropods did not rely heavily on their sense of smell, which probably isn't a major surprise. While sauropods had relatively small brains, their brain structure was likely sufficient to control their basic physiological needs, sensory processing, and perhaps minor motor functions. However, the large size of their spinal cords, especially in the pelvic region, suggests that much of their motor control for the hind limbs and tail was likely managed through the spinal cord rather than the brain.

All this to say that although sauropods weren't as advanced as say, a velociraptor, they likely did have an acute and highly evolved sense of instinct. This was a part of their well-adapted physiology, allowing their success and proliferation over 150 million years. We may also be thinking of intelligence in a conventional way; e.g., that cognition is solely confined to the brain. Recent research has proposed that intelligence is a collective property arising from the interactions of various cells, something called basal cognition. This burgeoning field challenges the notion that complex brains are prerequisites for intelligence and uses examples of sophisticated behavior in diverse life forms. The revelation that cells, beyond neurons, may possess memory storage capabilities has broad implications, with evidence suggesting that RNA itself could be a medium for memory in cell types. Clearly much more research needs to be done on this area of cellular behavior, memory, and intelligence, but it presents a possible mechanism whereby a new area of physiological research could be layered onto the sauropod, filling out an animal that may not have at first glance to have had a lot of "space between the ears."

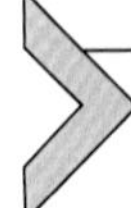

Did the sauropod even lift its head; that is, should the neck be raised?

After having reviewed the physiology and layered on a bit more with a small discussion on the sauropod central nervous system, it seems the right time to return to the point at the beginning of this chapter regarding the two evolutionary "design" components of the sauropod, its size and its neck. We noted that these were hardly flukes for the huge clad of sauropods to have been successful, but were an integral part to who they were as animals.

Originally, the sauropod was thought to be water-dwelling based mostly on the sheer size of the skeleton and neck length. This was then dispelled due to issues with the sauropod having no ballast to balance and thus an uneven center of gravity. There were other factors for this original hypothesis of water-dwelling sauropods being dismissed, including sauropod species radiation corresponding to areas of trees and savannahs, and the correlative ecological niches for eating along the height of a tree based on neck length. The biomechanics of water-dwelling were also an issue for the pneumatic-boned sauropod.

However, this debate has reemerged in the sauropod field about whether sauropods could lift their necks (similar to a giraffe), or if their necks were held in line with their heart (similar to a whale). The assumption has always been the sauropods lifted their necks, but invoking cardiovascular physiology, this has been challenged. The theory put forth that lifting the necks was not possible in the realm of concepts in cardiovascular physiology, was first put forward by Dr. Roger Seymour at University of Adelaide. As the theory goes, the blood pressure needed to push blood up the neck was simply too much, and the energy needed by the heart to produce those types of pressure was simply too great than the potential diet the sauropod was consuming. For that reason, Occam's razor suggested that the sauropod lived in water with its head held in line with the heart, similar to a whale. As noted in Chapters 4 and 5, this would keep blood pressure at a much lower level that was manageable for the sauropod. Besides reducing hydrostatic pressure, it allowed the heart to beat with less power and keep an overall lower blood pressure.

This theory cannot be dismissed for it follows strongly the requirements put forward by the physics behind the cardiovascular physiology. However, adaptions by the giraffe cardiovascular physiology suggest that animals can learn to cope with high pressures put on them by hydrostatic pressures of gravity. And as argued earlier, it is also possible that these cardiovascular adaptions seen in giraffes are scalable to sauropods, including hypertrophied left ventricles, a rete mirabile to dampen pressure as it approached the brain, and an overall much lower resistance in the vasculature to counteract the extremely high pressures. The great thing about new hypothesis is they challenge our thinking and lead to creative new ideas. Certainly more work in this area is warranted to explore this important hypothesis using physiology as our guide.

Why did they fall at the end of the Jurassic?

In the Mesozoic, there were several key climatic events that happened at the end of the Triassic period and beginning of the Jurassic period, mostly those with massive volcanic eruptions. This included the Central Atlantic

Magmatic and the Karoo-Ferrar volcanic provinces that had a dramatic effect on water and air temperatures. As previously mentioned, this coincided with an exponential rise in the variety of species of sauropods, which may have been due in large part to the warmer temperatures, increased CO_2 and thus vegetation allowing for the possible radiation of the species. What isn't as clear is why sauropods were so abundant during the Jurassic, but had a major crash from which they never really recovered at beginning of the Cretaceous. What about the sauropod physiology that could inform us of this possible downfall? Was their complete recovery possible if the Chicxulub impact had not occurred?

The sauropods were highly successful, but as the book has highlighted, highly specialized to their environment, making them susceptible to any kind of changes in the atmosphere. As we return to Copes law described in Chapter 2, their gigantism was a huge benefit to their survival, but with this size benefit came the downside of much greater risk in that they were so specialized to their environment, it wouldn't take major ecological events to place the sauropod continued radiancy into a tail spin. There is much work in this area looking at fluctuations in CO_2, temperature, and O_2 levels, and as models become more and more sophisticated to predict these events, perhaps these can give us some insight as we try to relate it back to their physiology. In effect, I suggest that the sauropods were a victim of their own success, and any little change in the environment in which their physiology had evolved could spell doom for the species.

The other possibility for the apparent population collapse is more logical and more simple—the large species radiation that drops off at the end of the Jurassic and beginning of the Cretaceous is simply due to the fact that more sauropod species have yet to be discovered that lived in the Cretaceous. Certainly, the titanosaurus clad of sauropods were quite successful in this time period. It's not the most fun explanation—lack of discovery—but with new dinosaur species discovered every day, it's something we can't dismiss.

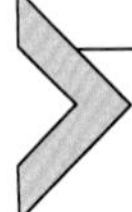

The many unknown unknowns of sauropod physiology...

We end the book the way we started the book, with the quote by former Secretary of Defense Donald Rummsfeld:

> *There are known knowns. These are things we know that we know. There are known unknowns. That is to say, there are things that we know we don't know. But there are also unknown unknowns. There are things we don't know we don't know.*

He was obviously not talking about sauropods at the time, he was trying to evade heated questions from reporters about why the United States was invading Iraq. Although it was initially thought to be just political jargon, the quote perfectly encapsulates what we know about dinosaurs in general, but sauropods especially.

For example, some known knowns about sauropods include that they had exceptionally long necks; they were massive in size with pneumatic bones; there were multiple species spread across earth; their fossilized feces show they ate leaves and conifers; they likely lived in herds; they had small brains at the top of their neck; and they lived in an environment with high CO_2, and low O_2.

There of course are a huge amount of known unknowns as well with sauropods, including the big one being how did they function in homeostasis as a complete animal with that long neck and giant size? How did their heart push enough blood up the neck? How mobile was that neck—was it always kept up, or was it truly always kept horizontal to minimize heart issues? Did the sauropod use avian-type lungs along their neck? Some of these questions I have tried to present some evidence for. However, these will likely remain known unknowns.

Finally, we arrive at the innumerable unknown unknowns about the sauropod's species of dinosaurs. Who knows what kind of discoveries are on the horizon that could change our view of how sauropods functioned? However, importantly, these discoveries will have to be made by paleontologists, not a physiologist. The reason for this, and the reason for even trying to make assumptions about sauropod physiology is because the hard evidence is derived from paleontological findings—this included what we know about bone size and structure, brain characteristics, and even the plants they ate and if their footprints were found together to deduce, they were in herds. We can only discuss the multiple hypotheses of the known unknowns above because the fossils give us the catalyst to imagine what their physiology was. The harder evidence we can get, the fuller picture we can deduce about what balanced the sauropod.

As we conclude, I've tried to provide essential physiology for the core parts of how a sauropod may have functioned and provided examples of how certain adaptions "tweak" the system depending on the environment in which the sauropod lived. The best we can do when trying to make assumption about how an extinct animal was in homeostasis with its environment is to try to layer our current knowledge of physiology on to the paleontological evidence provided for us. As more paleontological evidence is discovered about sauropods, my assumption is we can continually tweak our hypothesis about how sauropod physiology may have occurred. This will continually change, and that's not only ok and how science works, it's also exciting to think what will come next to challenge our ideas.

References and further reading

Aalkjær C, Wang T: The remarkable cardiovascular system of giraffes, *Annu Rev Physiol* 83:1–15, 2020.

Cope ED: The relation of animal motion to animal evolution, *Am Nat* 12:40–48, 1878.

Curry Rogers KA, Wilson JA, editors: *The sauropods*, Berkley and Los Angeles, California, 2005, University of California Press.

De Lena LF, Taylor D, Guex J, et al.: The driving mechanisms of the carbon cycle perturbations in the late Pliensbachian (early Jurassic), *Sci Rep* 9:18430, 2019.

Hallett M, Wedel MJ: *The sauropod dinosaurs*, Baltimore, Maryland, 2016, John Hopkins University Press.

Heimdal TH, Jones MT, Svensen HH: Thermogenic carbon release from the Central Atlantic magmatic province caused major end-Triassic carbon cycle perturbations, *Proc Natl Acad Sci USA* 117:11968–11974, 2020.

Klein N, Remes K, Gee CT, Sander PM, editors: *Biology of the sauropod dinosaurs*, Indianapolis, Indiana, 2011, Indiana University Press.

Prasad V, Stromber CE, Alimohammadian H, Sahni A: Dinosaur coprolites and the early evolution of grasses and grazers, *Science* 310:1177–1180, 2005.

Seifert G, Sealander A, Marzen S, Levin M: From reinforcement learning to agency: frameworks for understanding basal cognition, *Biosystems* 235, 2024 105107.

Seymour RS: Cardiovascular physiology of dinosaurs, *Physiology (Bethesda)* 31:430–441, 2016.

Wang X, Huang J, Hu Y, Liu X, Peteya J, Clarke JA: The earliest evidence for a supraorbital salt gland in dinosaurs in new early cretaceous ornighurines, *Sci Rep* 8:3969, 2018.

Wang X, Huang J, Hu Y, Liu X, Peteya J, Clarke JA: The earliest evidence for a supraorbital salt gland in dinosaurs in new early cretaceous ornithurines, *Sci Rep* 8:3969, 2018b.

Index

Note: Page numbers followed by *f* indicate figures.

Printed in the United States
by Baker & Taylor Publisher Services